Harvey Washington Wiley

Chemical Composition of the Carcasses of Pigs

Harvey Washington Wiley

Chemical Composition of the Carcasses of Pigs

ISBN/EAN: 9783337061685

Printed in Europe, USA, Canada, Australia, Japan

Cover: Foto ©berggeist007 / pixelio.de

More available books at **www.hansebooks.com**

Bulletin No. 53.

U. S. DEPARTMENT OF AGRICULTURE.
DIVISION OF CHEMISTRY.

CHEMICAL COMPOSITION

OF THE

CARCASSES OF PIGS.

BY

H. W. WILEY,
CHIEF OF THE DIVISION OF CHEMISTRY.

WITH THE COLLABORATION OF

E. E. EWELL, W. H. KRUG, T. C. TRESCOT, AND OTHERS.

WASHINGTON:
GOVERNMENT PRINTING OFFICE.
1898.

LETTER OF TRANSMITTAL.

U. S. DEPARTMENT OF AGRICULTURE,
DIVISION OF CHEMISTRY,
Washington, D. C., June 27, 1898.

SIR: I transmit herewith for your inspection and approval the manuscript containing the results of our investigations, undertaken at your suggestion, on the chemical composition of the carcasses of pigs grown at the agricultural experiment station of Iowa.

The scope of these investigations has extended so much farther than was at first anticipated as to render the results thereof worthy of publication as a separate bulletin of this Division. A study of the character of the data obtained will reveal at once their great importance, both from a scientific point of view and as a basis for economic studies.

The carcasses, as received by us, represented practically only those portions of the whole carcass which are subjects of commerce. The blood, hair, entrails, heads, kidneys, and kidney fats of the animals were removed before they were transmitted to us. The data, therefore, do not represent the composition of the whole animal, but what, perhaps, is of equal importance, the composition of the animal as sent into commerce for food.

In view of the great importance of investigations of this kind, I would venture to suggest that when the facilities for work in the chemical laboratories are extended by the completion of the new building now in course of construction, it would be well for you to direct that further studies of this kind be undertaken. It would be advisable, if possible, that in studies of this kind, the animals be slaughtered at or near the point where the chemical examination is to be made, or if this be not convenient, that a representative of the Chemical Division be present at the time of the slaughtering for the purpose of ascertaining the quantities of blood, hair, and excreta from the different animals and obtaining representative samples thereof for chemical examination.

I have the honor to be, respectfully,

H. W. WILEY,
Chief of Division.

Hon. JAMES WILSON, *Secretary.*

CONTENTS.

	Page.
Analytical work	7
Inception of the investigation	7
Correspondence	7
Breeds of hogs studied	8
Preparation of samples for analysis	9
Samples of meat	9
Samples of skin	10
Samples of bones and marrow	11
Samples of spinal cord	11
Samples of tendons	11
Samples of hoofs	11
Methods of analysis used	12
Results of the investigation	13
Description of tables	13
Tables	15–64
Discussion of the data	65
Composition of the same cuts from the different animals	65
Clear bellies	66
Short-cut hams	66
New York shoulders	67
Feet	67
Spareribs	68
Tenderloins	68
Neck bones	68
Backbones	68
Trimmings	69
Tails	69
Average of all cuts	69
Average of bones	70
Average of marrow	71
Average of skin	71
Average of spinal cord	73
Average of tendons	73
Average of hoofs	73
Loss of weight in transportation	74
Ratios of meat, bones, etc., to total weight	74
Percentages of the several constituents	75
Comparison of breeds	75
Lecithin	76
Physiological importance	76
Discussion of the lecithin in particular samples	77
Lecithin in the meat	77
Lecithin in the bones	77
Lecithin in the marrow	77
Lecithin in the skins	77
Lecithin in the spinal cord	77
Lecithin in the tendons	77

Concluding observations	78
Appendix	79
Precipitation of proteids soluble in water by chlorin and bromin	79
Nitrogen in meat extracts	80
Problems solved by the bromin method	80

CHEMICAL COMPOSITION OF THE CARCASSES OF PIGS.

ANALYTICAL WORK.

INCEPTION OF THE INVESTIGATION.

Following instructions received from the Secretary of Agriculture, the Division of Chemistry, in November, 1897, undertook a study of the chemical composition of the carcasses of pigs. These pigs were grown at the Iowa Agricultural Experiment Station under standard conditions of diet, and a comparison of their carcasses reveals, therefore, the influence of breed and heredity on the character of the meat. In the following correspondence will be found the data connected with the history of the animals before they were delivered to the Division of Chemistry.

CORRESPONDENCE.

EXPERIMENT STATION, IOWA AGRICULTURAL COLLEGE,
Ames, Iowa, October 8, 1897.

MY DEAR SIR: We have, as you are aware, a very interesting and instructive experiment nearing completion, in which we have grown carefully selected representatives of six of the leading breeds of hogs since birth in lots of ten each. These pigs are now weighing nearly 200 pounds, and will be forwarded to market for the test in determining the relative market value and the results in slaughtering and on the block, and the meat will be carefully compared and rated by experts. This experiment includes the Poland China, Berkshire, Duroc Jersey, Chester White, Tamworth, and Yorkshire. It has occurred to us that a careful and exhaustive chemical analysis of representative carcasses selected from each lot after slaughtering would be a valuable feature of this investigation, and I write to know if the Department of Agriculture can not cooperate with us in this work. We will gladly furnish you such material as may be needed and in any form desired. I will be glad to hear from you in reference to this point, and trusting that such arrangements can be made, I am,

Very truly, yours,

C. F. CURTISS.

Hon. JAMES WILSON,
Secretary of Agriculture, Washington, D. C.

EXPERIMENT STATION, IOWA AGRICULTURAL COLLEGE,
Ames, Iowa, October 29, 1897.

MY DEAR Mr. WILSON: Your esteemed favor of the 26th instant is at hand and I note what you say about cooperation of the Department with us in our hog-feeding experiments. The final weighing of the pigs will be taken Monday, and they will arrive in Chicago Tuesday morning. I have arranged to place them on exhibition in the Coliseum Building during the fat-stock show, and will take them to the stock yards for slaughter and block tests immediately following. After the carcasses have been cooled down I will have a committee of the expert meat dealers select one or two representative carcasses from each lot and forward to Dr. Wiley for investigation. Probably it will be a week or ten days before the carcasses reach Washington.

Very truly, yours,

C. F. CURTISS.

Hon. JAMES WILSON,
   Secretary of Agriculture, Washington, D. C.

EXPERIMENT STATION, IOWA AGRICULTURAL COLLEGE,
Ames, Iowa, November 13, 1897.

DEAR SIR: Your valued favor of the 3d instant came to hand while I was in Chicago having the slaughter test made of the pigs used in our experiments. Owing to the machinery used in the packing house where the hogs were killed, it was not practicable to obtain the weight of the hair, and the blood could not be collected and weighed without considerable difficulty. I had taken this matter up with Swift & Co. before receiving your letter, but was obliged, under the circumstances, to omit these items. The weight of the intestines and other internal organs was obtained. I returned this morning from superintending the block test yesterday, and have had a good representative carcass from each lot selected and cut according to the prevailing method of cutting pork for the American market, and each piece weighed and properly tagged, giving commercial names. I think, however, that the names are appended only to one set of cuts, but you will be able to apply these names to corresponding cuts of the other carcasses. I have directed Swift & Co. to forward this material to you, including all scraps and trimmings made in cutting, and to deliver it to you at their earliest convenience. They stated that they would probably have one of their refrigerator cars leaving for Washington to-day, and that they would notify you upon its arrival at their house in Washington and deliver the pork upon your order.

Very truly, yours,

C. F. CURTISS.

Dr. H. W. WILEY,
   Chief of Division of Chemistry, Washington, D. C.

BREEDS OF HOGS STUDIED.

In accordance with the plan outlined in the above letters, on November 16, 1897, Swift & Co., of Chicago, shipped to the Department of Agriculture the carcasses of eight pigs which had been slaughtered under the direction of Professor Curtiss. These pigs were of the following breeds, each animal being designated by a number, which is used for its identification throughout the following pages:

1, Berkshire; 2, Tamworth; 3, Chester White; 4, Poland China; 5, Duroc Jersey; 6, Duroc Jersey; 7, Duroc Jersey; 8, Yorkshire.

On the receipt of the animals in Washington, they were immediately placed in cold storage, where they were kept until they were removed one by one for the purpose of dissecting and preparing the samples for analysis.

The expert labor of assistants in the meat markets of Washington was secured for the purpose of properly dissecting the animals and separating each portion as carefully as possible from the others. The greatest care was exercised in this preliminary work, inasmuch as the value of the analytical data rests largely on the proper preparation of the materials for examination.

PREPARATION OF SAMPLES FOR ANALYSIS.

The methods of preliminary treatment, together with the methods of chemical analysis employed, are detailed in the following pages. Before leaving Chicago each animal was cut up into the following cuts, the head, leaf lard, and kidneys being retained in Chicago:

Two American clear backs; two clear bellies; two short-cut hams; two New York shoulders; four feet; spare ribs; tenderloins; neck bones; back bones; trimmings, fat and lean; tail.

These cuts were all weighed on leaving Chicago, and again in Washington just preceding their analysis. All of these weights appear in the accompanying tables, pages 15 to 64. The weighings in Washington were made on a large counter scale for the larger cuts, and on a torsion balance in the case of the smaller cuts. The cuts were then separated into the following parts: Meat (including both fat and lean), bones, marrow, skin, spinal cord, tendons, and hoofs.

Each of the parts, except the meat, was carefully weighed, and the weight of the meat obtained by subtracting the sum of the other weights from the total weight of the cut before cutting up.

SAMPLES OF MEAT.

The meat obtained from all of the cuts of the same kind in each animal was passed through a meat chopper two or more times in order to bring the sample into a finely divided condition. A weighed portion was then placed in a weighed casserole or evaporating dish. A glass rod was also weighed with the casserole. In the case of small samples, as the tenderloins, the entire quantity was taken; in the case of the larger cuts, from 400 to 600 grams of the fresh material were taken for the preparation of the air-dried sample. After the removal of these portions for the preparation of the air-dried sample, duplicate portions of 5 grams each were weighed for the direct determination of water and fat. These small samples were placed in aluminum dishes and dried in vacuo for six hours at 105 degrees. The residues were extracted for sixteen hours with ether, and the extracts dried in an air bath at 100 degrees. These direct determinations of fat and water were used as a check on the data obtained in the preparation of the air-dry samples. The larger portions, which had been weighed out as described above for the preparation of the air-dried samples, were placed in a steam oven at a temperature of 100 degrees or slightly more and heated until the fat had thoroughly separated, when the fat was poured off into a flask, care being taken not to pour with it any of the aqueous portion of the meat which formed a layer underneath the fat. After as much fat had been poured off as was possible, the drying was continued in the steam oven until the weight had become approximately constant. As there was still too much fat contained in the samples to permit of their being powdered, it was necessary to extract them with ether before proceeding with the grinding. The extraction with ether was done in the following way.

Large funnels were placed in hot-water jackets, and in the funnels were placed filters of parchmentized paper. The smooth surface of this paper greatly facilitated the removal of the insoluble residue of the sample. The portion of fat from each sample, which had been poured off as above described, was first passed through this filter and collected in a weighed flask and its weight taken. The remainder of the sample was then treated with ether and brought on to the filter and the washing with ether continued until the fat was sufficiently removed for the sample to be easily pulverized and brought into proper condition for subsequent analytical operations. The ether solution of the fat was also received in a weighed flask. The ether was removed by distillation and the residue heated to constant weight and weighed. There was considerable annoyance from the breaking of the flasks containing the fat while on the steam bath. When there was an evident loss of fat, the fat determinations were recorded as lost. When the flask was discovered with only a slight crack, the results are marked in the following tables with a (?) mark. The portion of the meat on the filter was returned to the dish which had previously contained it, and was again dried to approximately constant weight and then left exposed to the air for at least twenty-four hours in order to establish an equilibrium of its moisture content. The weight of the sample was then taken and recorded as the air-dry weight of the material.

The difference obtained by subtracting the sum of the weights of the air-dry material, fat obtained by pouring, and fat obtained by ether extraction from the original weight of the sample taken was recorded as the weight of water removed in the preparation of the sample. From these data were calculated:

Percentage of water removed in the preparation of the sample;
Percentage of fat removed in the preparation of the sample; and
Percentage of air-dry sample obtained.

All three of these were expressed in percentages of the original material.

The air-dry samples were then ground, so as to pass a sieve having circular perforations 1 millimeter in diameter, and placed in closely stoppered bottles.

SAMPLES OF SKIN.

The portions of skin obtained from each cut were united to make one sample of skin for the entire animal. The united sample of skin from each pig was passed through the meat chopper, and the finely divided and thoroughly mixed sample was treated in exactly the same way as described above for the samples of meat. The samples of meat from each cut were kept separate, however, while only one sample of skin was prepared for each animal.

SAMPLES OF BONES AND OF MARROW.

The bones from each cut were weighed and were united to make one sample of bones from each animal. They were then chopped up into bits about 1 inch long and the marrow removed. The marrow was weighed in a tared dish and treated as samples of meat, except that no determinations of moisture and fat were made in the original material. The fragments of the bones after the removal of the marrow were thoroughly mixed, and about half the total quantity was weighed in a tared dish and dried to approximately constant weight in a large agate-ware pan. After standing for from twenty-four to forty-eight hours exposed to the air, the weight was again taken and recorded as the weight of air-dried bones equivalent to the portion of fresh bones taken for the drying. The sample thus obtained was passed through a bone cutter, such as is used for poultry food, and from this, 500-gram portions were weighed and treated with petroleum ether by decantation for the removal of the fat. The solutions of fat were very difficult of filtration, hence were allowed to stand for some time for the almost complete subsidence of the solid matter contained in them, when they were carefully siphoned off and evaporated and the weight of the fat contained in them determined. The residues were again dried and exposed to the air for the establishment of the equilibrium of moisture content, and again weighed, the weight obtained being recorded as the weight of the air-dry, extracted bones. The samples thus obtained were submitted to analysis, and the determinations made are recorded below, all percentages being calculated back to the original material by use of the data obtained in the preparation of the sample.

SAMPLES OF SPINAL CORD.

The spinal cord was carefully separated from the backbones and neck bones, and the material thus obtained united to make one sample of spinal cord for each animal. This sample was prepared for analysis in the manner described for meats, but it was not practicable to make a direct determination of fat and moisture in the original sample.

SAMPLES OF TENDONS.

It was not practicable to separate the tendons from other cuts of the animal than the feet and legs, that is, the portion sent to the laboratory under the name of "feet." The tendons were treated in the same manner as the spinal cord.

SAMPLES OF HOOFS.

The hoofs were separated and weighed. In some cases some of the hoofs had been removed in the process of slaughtering and dressing the animal. In these cases the whole weight of hoofs was corrected for the deficiency by using the average weight of one hoof for the weight of each of the remaining hoofs. The hoofs were weighed and dried in the steam oven and then left to assume their air-dry content of moisture. They were then ground and submitted to analysis as described below for the other parts.

METHODS OF ANALYSIS USED.

On the samples thus prepared the following determinations were made:

Water, fat, ash, total nitrogen, nitrogen insoluble in hot water, nitrogen soluble in hot water but precipitated by bromin, and lecithin.

For the determination of MOISTURE and FAT 2-gram portions were dried for six hours in a vacuum oven for the determination of water, and the residues were extracted for sixteen hours with ether for determination of the fat.

For TOTAL NITROGEN duplicate portions of one-half gram of the air-dried sample were treated by the Gunning method.

For INSOLUBLE PROTEID NITROGEN 1-gram portions were washed with ether by decantation, using about 50 to 100 c. c. of ether for each sample, and decanting the ether through filters which were afterwards used to receive the portions of the sample insoluble in hot water. After allowing the ether to evaporate the samples were next treated with hot water, this washing being also by decantation, and the total amount of water used being 300 to 400 c. c., the residues being brought on the filter with the last portion of the water. The filters and residues were then treated by the Gunning method.

The filtrates from the insoluble portions of the meat were received in Kjeldahl flasks and were used for the determination of the NITROGEN PRECIPITATED BY BROMIN (GELATINOIDS).[1] After acidulation with two or three drops of strong hydrochloric acid, about 2 c. c. of bromin were added and the flasks vigorously shaken. If this quantity of bromin was all taken up more was added and the shaking repeated until a globule of about $\frac{1}{2}$ c. c. of bromin was left in the flask, and the liquid above it was thoroughly saturated with bromin. The mixture was then allowed to stand until the next morning, when the supernatant liquor was passed through a filter and the residue in the flask washed by decantation, the globule of undissolved bromin in the flask saturating the wash water with bromin, so that it was unnecessary to use bromin water for the washing. The filter containing the residue was then returned to the same flask in which the precipitation had taken place and treated by the Gunning method.

The percentage of nitrogen in the form of FLESH BASES was found by subtracting the sum of the numbers representing insoluble nitrogen

---

[1] See Appendix, page 79.

and nitrogen precipitated by bromin from the number representing the percentage of total nitrogen. The percentage of flesh bases was obtained by multiplying the percentage of nitrogen in that form by 3.12. For the other forms of nitrogen, the factor 6.25 was used.

For the determination of LECITHIN,[1] 20 grams of the material were allowed to stand for twenty-four hours at from 35° to 40° C. with 200 c. c. of a mixture of equal parts of ether and 95 per cent alcohol. The material was then filtered and the residue extracted repeatedly with the same solvent. The filtrate and washings were evaporated to dryness on the water bath in a platinum dish. The residue was fused with mixed carbonates (equal parts of sodium and potassium carbonates). A little potassium nitrate was added during the fusion. The flux was dissolved in hot water, filtered, and the phosphoric acid determined in the filtrate by the Kilgore-Pemberton volumetric method. The lecithin was calculated as distearyl lecithin, which contains 8.789 per cent P_2O_5.

RESULTS OF THE INVESTIGATION.

DESCRIPTION OF TABLES.

The results of this work are presented in the accompanying tables. The first fifty-six tables are in seven groups. Each group gives in a separate table data for each of the eight pigs used.

Table 1 shows the weights of the whole cuts as obtained in Chicago and Washington, results of the direct determination of water and fat in the meat from each cut, and data in regard to the preparation of the air-dry sample of the meat from each cut.

Tables 2 and 3 show the weights of meat, bones, skin, etc., obtained from each cut, the total for the whole animal, and also the percentages of meat, bones, skin, etc., in each animal. These sheets also contain the data in regard to the preparation of the sample of bones, marrow, skin, spinal cord, tendons, and hoofs.

Tables 4 and 5 show all the analytical data, including the data actually obtained on the air-dry material, and also the corresponding data expressed in terms of the original material.

In Table 6 the analytical data have been collected in condensed form for convenience of reference.

In Table 7 are presented the weights of water, fat, nitrogenous substances, lecithin and ash in the meat of each entire animal, and also the weights and average percentages of each of these substances for the entire animal, including all its parts—meat, bones, skin, etc. These data were obtained by multiplying the weight of the meat from each cut by the percentage of each one of the constituents, finding the total, and dividing by the number representing the total weight of the meat of the entire animal. The same method was employed for the bones, marrow,

---

[1] Principles and Practice of Agricultural Analysis, Vol. III, p. 430.

skin, etc. Thus there were obtained the total weight of water for each animal, total weight of fat, etc. These total weights, divided by the weight of the entire animal, gave the average percentages of the various constituents of the entire animal.

In Tables 8 A to 8 K, have been placed the data which show the chemical composition of the meat of each cut of each pig.

In Table 9 has been placed the average composition of the meat of each animal.

Table 10 contains similar data for the bones of each animal, Table 11 for the marrow, Table 12 for the skin, Table 13 for the spinal cord, Table 14 for the tendons, and Table 15 for the hoofs. In Table 16 will be found a résumé of the weights of each cut, and also of each entire animal, as found in Chicago and found in Washington, the results being stated in both grams and pounds.

Table 17 shows the percentages of each of the parts for each animal, stated in percentages of the entire dressed animal, less the head, leaf lard, and kidneys.

Table 18 shows the proportion of water, fat, nitrogenous substances, lecithin and ash in each of these animals, stated in percentages of the entire dressed animal, less the head, leaf lard, and kidneys.

There is one obvious omission in the data presented in the tabulation just described. The absence of any information in regard to the manner of the feeding of the pigs has made it impossible to group them properly and make proper averages of the percentages of the various constituents in the animals which have received the same rations and other treatment previous to their slaughter. This missing data will be found in the forthcoming full report of Professor Curtiss, of the Iowa Agricultural Experiment Station, which should be consulted with the data herewith submitted.

15

TABLE No. 1.—*Weights of whole cuts and data relating to the preparation of air-dry samples.*

PIG No. 1.—BERKSHIRE.

Serial No. of air-dry meats.	Names of cuts.	Weights of whole cuts.				Direct determinations on original material.		Weight of fresh sample.	Air-dry sample of original material.	Preparation of air-dry samples.				Removed in preparation of sample.	
		Chicago.		Washington.		Water.	Fat.			Weight of air-dry sample after extraction.	Weight of fat.	Air-dry sample plus fat.	Weight of water removed.	Water.	Fat.
		Lbs. Oz.	Grams.	Lbs. Oz.	Grams.	Per ct.	Per ct.	Grams.	Per ct.	Grams.	Grams.	Grams.	Grams.	Per ct.	Per ct.
16067	2 American clear backs	35½ 0	16,102.8	34 0	15,592.5										
	Meat					31.33	58.21	833.0	13.16	109.6	458.0	567.6	265.4	31.86	54.98
16069	2 clear bellies	19½ 0	8,845.2	19 4	8,731.8										
	Meat					36.09	52.69	741.2	14.33	106.2	362.1	468.3	272.9	36.82	48.84
16071	2 short-cut hams	23½ 0	10,659.6	23 5	10,574.6										
	Meat					60.29	22.19	532.5	22.95	122.2	88.3	210.5	322.0	60.47	16.58
16073	2 New York shoulders	20½ 0	9,298.8	20 10	9,395.5										
	Meat					54.97	29.01	532.5	17.65	94.0	152.9	246.9	285.6	53.64	28.71
16075	4 feet (7 hoofs)	a 3½ 0	b1,594.2		b1,614.1										
	Meat					59.78	17.04	221.1	25.10	55.5	33.7	89.2	131.9	59.60	15.24
16077	Spareribs	5 0	2,268.0		2,212.0										
	Meat					50.33	30.05	359.9	20.81	74.9	98.6	173.5	186.4	51.74	27.39
16079	Tenderloins	1 0	453.6		470.8	67.11	9.14	427.9	27.11	116.0	26.8	142.6	285.3	66.07	6.21
	Neck bones	2 0	907.2		842.5										
	Meat					53.82	28.72	390.6	20.02	78.2	100.5	178.7	211.9	54.25	25.73
16080	Backbones	3½ 0	1,587.6		1,580.0										
	Meat					51.89	27.16	397.5	22.24	88.4	102.1	190.5	207.0	52.08	25.69
16082	Trimmings	18 0	8,164.8	16 9	7,512.8										
	Meat					29.68	62.00	783.7	9.72	76.3	479.2	555.5	228.2	29.11	61.17
16084	Tail	¼ 0	113.4		363.0										
	Meat					23.99	69.25	199.2	8.73	17.4	134.7	152.1	47.1	23.04	67.62
16086	Total	132¼ 0	b59,995.2		58,789.6										

a Missing hoof, 6.5 grams.   b Corrected for missing hoof.

TABLE No. 1.—*Weights of whole cuts and data relating to the preparation of air-dry samples*—Continued.

PIG No. 2.—TAMWORTH.

Serial No. of air-dry meats.	Names of cuts.	Weights of whole cuts.				Direct determinations on original material		Preparation of air-dry samples.						Removed in preparation of sample.	
		Chicago.		Washington.		Water.	Fat.	Weight of fresh sample.	Air-dry sample of original material.	Weight of air-dry sample after extraction.	Weight of fat.	Air-dry sample plus fat.	Weight of water removed.	Water.	Fat.
		Lbs.	*Oz.*	*Grams.*	*Lbs.*	*Oz.*	*Grams.*	*Per ct.*	*Per ct.*	*Grams.*	*Per ct.*	*Grams.*	*Grams.*	*Grams.*	*Grams.*	*Per ct.*	*Per ct.*
16696	2 American clear backs Meat	41	0	18,597.6	40	8	18,370.8	29.13	61.76	786.0	10.48	82.4	377.6	460.0	326.0	41.48	48.04
16698	2 clear bellies Meat	20	0	9,072.0	19	9	8,873.6	32.74	58.00	779.4	12.12	94.5	426.4	520.9	258.5	33.17	54.71
16700	2 short-cut hams Meat	26	0	11,793.6	25	12	11,680.2	57.93	24.45	617.4	19.99	122.9	Lost.				
16702	2 New York shoulders Meat	21	0	9,525.6	20	12	9,412.2	35.07	29.98	827.6	20.36	168.5	Lost.				
16704	4 feet (8 hoofs) Meat	a4¼	0	b2,057.3			b1,974.1	58.86	21.23	331.7	23.85	79.1	58.9	138.0	193.7	58.39	17.76
16706	Spareribs Meat	5	0	2,268.0			2,132.7	49.58	19.79	701.0	19.67	137.9	225.8	363.7	337.3	48.12	32.21
16708	Tenderloins Meat	1	0	453.6			528.2	63.00	13.71	420.3	24.12	101.4	48.3	149.7	270.6	84.38	11.49
16709	Neck bones Meat	2	0	907.2			886.0	54.21	27.14	322.7	20.95	67.6	79.6	147.2	175.5	54.38	24.67
16711	Backbones Meat	4	0	1,814.4			1,840.0	50.45	30.33	423.3	20.98	88.8	124.2	213.0	210.3	49.66	29.34
16713	Trimmings Meat	18¼	0	8,278.2	16	10	7,541.1	27.39	66.33	869.8	10.53	91.6	531.7	623.3	246.5	29.34	61.13
16715	Tail Meat	¼	0	113.4			707.5	29.38	61.54	474.2	9.53	45.2	120.0	165.2	309.0	25.31	65.16
	Total	143	0	b64,880.9			63,946.4										

PIG No. 3.—CHESTER WHITE.

16609	2 American clear backs Meat	36	0	16,329.6	35	12	16,216.2	23.72	70.16	1,094.3	10.09	110.5	(?)197.2	307.7	786.6	71.88	18.02
16611	2 clear bellies Meat	21	0	9,525.6	21	0	9,525.6	29.40	62.23	1,027.0	9.81	100.7	614.9	715.6	311.4	30.32	50.87
16613	2 short-cut hams Meat	20	0	9,072.0	19	15¼	9,057.8	53.41	29.09	628.4	24.30	152.7	145.7	298.4	330.0	52.51	23.19
16615	2 New York shoulders Meat	21	0	9,525.6	20	13	9,440.5	49.16	37.62	762.2	17.75	135.3	Lost.				

	4 feet (4 hoofs)	c2½	0	b1,152.5			b1,236.9										
16617	Meat							53.05	20.74	144.4	20.71	29.9	37.5				
	Spareribs	3	0	1,360.8			1,409.0										
16619	Meat							53.46	27.14	478.8	26.69	127.8	99.9	227.7	251.1	52.44	30.87
16621	Tenderloins	1	0	453.6			453.6	66.75	11.97	435.5	25.88	112.7	39.9	152.6	282.9	64.96	9.16
	Neck bones	1½	0	680.4			683.7										
16622	Meat							54.19	27.04	347.1	19.30	67.0	96.3	163.3	183.8	52.96	27.74
	Backbones	2½	0	1,134.0			1,172.6										
16624	Meat							50.82	31.03	481.3	19.39	(93.3)	· 147.2	240.5	240.8	50.03	30.58
	Trimmings	27	0	12,247.2	15	12	7,144.2										
16626	Meat							19.17	76.04	992.5	7.20	71.5	(?) 700.6	772.1	220.4	22.21	70.59
	Tail	¼	0	113.4			740.2										
16628	Meat							15.66	79.69	553.9	5.65	31.3	Lost.				
	Total	135¼	0	b61,594.7			57,080.3										

PIG No. 4.—POLAND CHINA.

	2 American clear backs	40	0	18,144.0	38	12	17,577.0										
16638	Meat							26.13	66.33	954.0	9.68	92.3	(?) 582.9	675.2	278.8	29.22	61.10
	2 clear bellies	24	0	10,886.2	23	10	10,716.1										
16640	Meat							29.22	62.65	820.5	9.43	77.4	493.8	571.2	249.3	30.38	60.19
	2 short-cut hams	26	0	11,793.6	25	14	11,736.9										
16642	Meat							53.14	31.32	543.1	17.00	92.3	157.7	250.0	293.1	53.97	29.03
	2 New York shoulders	24	0	10,886.2	23	7	10,631.1										
16644	Meat							51.72	33.74	557.5		103.7	Lost.				
	4 feet (8 hoofs)	3	0	1,360.8			1,359.0										
16646	Meat							47.97	33.32	203.7	19.57	39.8	63.6	103.4	100.3	49.23	31.20
	Spareribs	5	0	2,268.0			1,969.5										
16648	Meat							52.95	29.55	503.3	22.03	110.9	(?) 112.9	223.8	279.5	55.53	22.44
16650	Tenderloins	1	0	453.6			419.8	66.88	11.58	392.9	26.80	105.3	28.9	134.2	258.7	65.85	7.36
	Neck bones	1½	0	680.4			815.6										
16651	Meat							53.74	30.08	432.1	19.37	88.7	114.7	198.4	233.7	54.08	26.55
	Backbones	3	0	1,360.8			1,315.5										
16653	Meat							51.26	30.98	452.2	21.21	95.9	(?) 136.3	232.2	220.0	48.65	30.14
	Trimmings	21¼	0	9,639.0	20	0	9,072.0										
16655	Meat							21.96	72.19	950.0	8.05	76.5	- 663.2	739.7	210.3	22.14	69.81
	Tail	¼	0	113.4			760.0										
16656	Meat							16.56	78.83	570.0	6.41	36.5	441.5	478.0	92.0	16.15	77.44
	Total	149	0	67,586.4			66,372.5										

a Missing hoofs, 16.1 grams.

b Corrected for missing hoofs.

c Missing hoofs, 18.5 grams.

TABLE No. 1.—*Weights of whole cuts and data relating to the preparation of air-dry samples*—Continued.

PIG No. 5.—DUROC JERSEY.

Serial No. of air-dry meats.	Names of cuts.	Weights of whole cuts.				Direct determinations on original material.		Preparation of air-dry samples.					Removed in preparation of sample.	
		Chicago.		Washington.		Water.	Fat.	Weight of fresh sample.	Air-dry sample of original material.	Weight of air-dry sample after extraction.	Weight of fat.	Air-dry sample plus fat.	Weight of water removed.	Water.	Fat.
		Lbs. Oz.	Grams.	Lbs. Oz.	Grams.	Per ct.	Per ct.	Grams.	Per ct.	Grams.	Grams.	Grams.	Grams.	Per ct.	Per ct.
16379	2 American clear backs	39½ 0	17,917.2	39½ 0	17,917.2	20.75	73.25	1,086.1	9.37	101.8	Lost.				
	Meat														
16581	2 clear bellies	24 0	10,886.4	24½ 0	10,943.1	28.35	64.67	969.5	11.71	115.8	589.1	704.9	284.6	28.77	59.52
	Meat														
16583	2 short-cut hams	21 0	9,525.6	21 0	9,525.6	49.57	35.73	1,213.3	21.39	259.5	350.7	610.2	603.1	49.71	28.90
	Meat														
16585	2 New York shoulders	19½ 0	8,845.2	19 13	8,987.0	44.16	43.74	703.3			293.1			41.67	
	Meat														
16587	4 feet (7 hoofs)	a 2½ 0	b 1,137.0		b 1,255.4	54.16	26.19	102.3	23.36	23.9	23.8	47.7	54.6	53.37	23.27
	Meat														
16589	Spareribs	3½ 0	1,587.6		1,612.0	53.80	27.08	604.7	22.23	134.4	147.5	281.9	322.8	53.38	24.39
16591	Tenderloins	1 0	453.6		348.5	67.39	10.55	315.0	28.54	89.9	19.3	109.2	205.8	65.34	6.12
	Neck bones	1½ 0	680.4		784.0										
16592	Meat	3 0	1,360.8		1,438.0	52.83	29.61	395.8	19.46	77.0	114.1	191.1	204.7	51.71	28.83
16594	Backbones														
	Meat					52.06	29.47	577.5	25.94	149.8	151.2	301.0	276.5	47.88	26.18
16596	Trimmings	20½ 0	9,185.4	19 11	8,930.3	19.57	74.46	979.2	6.80	66.5	718.7	785.2	194.0	19.81	73.39
	Meat														
16598	Tail	½ 0	113.4		663.0	11.13	85.20	532.8	4.70	25.0	447.4	472.4	60.4	11.33	83.97
	Meat														
	Total	136 0	61,693.5		62,424.1										

PIG No. 6.—DUROC JERSEY.

16725	2 American clear backs	45 0	20,412.0	44 8	20,185.2	20.74	74.09	740.1	6.99	51.7	541.5	593.2	146.9	19.85	73.16
	Meat														
16727	2 clear bellies	32½ 0	14,742.0	32 11	14,827.1	2.41	72.20	793.0	8.30	65.8	456.7	522.5	270.5	34.11	57.59
	Meat														
16729	2 short-cut hams	27 0	12,247.2	26 14	12,190.5	42.08	45.54	671.3	27.63	185.5	245.9	431.4	239.9	35.74	36.63
	Meat														
16731	2 New York shoulders	22 0	9,979.2	22 2	10,035.9	38.62	50.66	544.6	12.91	70.3	264.0	334.3	210.3	38.61	48.48
	Meat														

16733	4 feet (8 hoofs)	3½	0	1,587.6		1,547.2	49.65	31.99				
	Meat								292.1	21.84	63.8	81.8	145.6	146.5	50.16	28.00
16735	Spareribs	4	0	1,814.4		1,612.2	48.50	33.45	490.7	22.09	108.4	144.0	252.4	238.3	46.56	29.35
16737	Tenderloins	½	0	226.8		421.3	62.50	14.91	377.6	26.77	101.1	41.1	142.2	235.4	62.34	10.89
	Meat	2	0	907.2		892.6	48.39	35.99	476.7	17.37	82.8	163.9	246.7	230.0	48.25	34.38
16738	Neck bones															
	Meat	3½	0	1,587.6		1,546.0	47.54	35.96	487.1	21.13	102.9	257.1	360.0	127.1	26.09	52.78
16740	Backbones															
	Meat	27¼	0	12,360.6	25	1	11,368.4	16.45	81.93	678.0	5.85	39.7	527.9	567.6	110.4	16.28	77.87
16742	Trimmings															
16744	Tail	¼	0	113.4		1,173.0	14.24	82.55	742.3	6.37	47.3	591.5	638.8	103.5	13.93	79.70
	Meat															
	Total	167½	0	75,978.0		75,799.4										

PIG No. 7.—DUROC JERSEY.

16754	2 American clear backs	39½	0	17,917.2	38	14	17,633.7	19.20	76.97	804.5	7.39	59.4	(?) 587.5	646.9	157.6	19.59	73.03
	Meat															
16756	2 clear bellies	28½	0	12,927.6	29	0	13,154.4	21.53	73.56	744.7		57.6	(?) 789.9				
	Meat															
16758	2 short-cut hams	23½	0	10,659.6	23	13	10,801.4	44.26	43.38	578.3	14.73	85.2	245.2	330.4	247.9	42.87	42.40
	Meat															
16760	2 New York shoulders	19½	0	8,845.2	19	12	8,938.6	44.60	43.56	641.5	14.90	95.6	265.6	301.2	280.3	43.68	41.42
	Meat															
16762	4 feet (1 hoof)	c2¼	0	b1,134.0		b1,400.0	53.76	26.43	215.2	21.46	46.2	51.4	97.6	117.6	54.66	23.88
	Meat															
	Spareribs	3½	0	1,587.6		1,504.0	52.86	27.52	484.7	21.77	105.5	127.1	232.6	252.1	52.01	26.22
16764	Meat															
16766	Tenderloins	2	0	340.2		333.3	65.70	11.57	194.15	20.76	40.3	14.2	54.5	139.65	71.93	7.31
	Meat	2	0	907.2		906.5										
16767	Neck bones															
	Meat	3	0	1,360.8		1,482.0	50.02	33.30	224.5	20.09	45.1	65.0	110.1	114.4	50.96	28.95
16769	Backbones															
	Meat	24½	0	11,226.6	23	11	10,744.7	51.75	30.71	231.2	22.45	51.9	59.1	110.0	120.2	51.99	25.56
16771	Trimmings															
16773	Tail	¼	0	113.4		759.0	21.30	73.41	866.3	7.35	63.8	630.3	694.1	174.2	20.06	72.59
	Meat															
	Total	147½	0	b67,019.4		b67,677.6	13.22	85.04	474.5	5.94	28.2	382.6	410.8	63.7	13.43	80.63

a Missing hoof, 3.9 grams.   b Corrected for missing hoofs.   c Missing hoofs, 45.5 grams.

TABLE No. 1.—*Weights of whole cuts and data relating to the preparation of air-dry samples*—Continued.

PIG No. 8.—YORKSHIRE.

Serial No. of air-dry meats.	Names of cuts.	Weights of whole cuts.				Direct determinations on original material.		Preparation of air-dry samples.					Removed in preparation of sample.	
		Chicago.		Washington.		Water.	Fat.	Weight of fresh sample.	Air-dry sample of original material.	Weight of air-dry sample after extraction.	Weight of fat.	Air-dry sample plus fat.	Weight of water removed.	Water.	Fat.
		Lbs. Oz.	Grams.	Lbs. Oz.	Grams.	Per ct.	Per ct.	Grams.	Per ct.	Grams.	Grams.	Grams.	Grams.	Per ct.	Per ct.
16783	2 American clear backs	44 0	19,958.4	43 13	19,873.4										
	Meat					28.00	63.88	884.2	9.67	85.45	551.2	636.7	247.5	27.99	62.34
16785	2 clear bellies	22½	10,206.0	23 1	10,461.2										
	Meat					33.58	36.97	926.8	11.09	102.8	517.5	620.3	306.5	33.08	55.83
16787	2 short-cut hams	27 0	12,247.2	27 4	12,380.6										
	Meat					54.80	28.64	704.7	21.97	154.8	141.1	295.9	408.8	58.02	20.01
16789	2 New York shoulders	24½	11,113.2	25 5	12,502.4										
	Meat					49.57	36.09	646.8		133.8	Lost.				
16791	4 feet (8 hoofs)	4½	2,041.2	4 14	2,246.0										
	Meat					57.47	30.86	284.7		62.0	Lost.				
16793	Spareribs	5 0	2,268.0	5 2	2,340.0										
	Meat					51.55	30.18	561.4	22.26	125.0	150.5	275.5	285.9	50.93	26.81
16795	Tenderloins	1 0	453.6	1 6	632.5										
	Meat					65.24	13.35	461.3	25.23	116.4	43.0	159.4	301.9	65.45	9.32
16796	Neck bones	2 0	907.2	2 10	1,192.3										
	Meat					53.25	29.29	439.8	19.92	87.6	115.3	202.9	236.9	53.87	26.21
16798	Backbones	4 0	1,814.4	4 6	1,998.0										
	Meat					49.84	30.58	727.8	23.55	171.4	197.4	368.8	359.0	49.33	27.12
16800	Trimmings	24¾	11,226.6	18 10	8,448.3										
	Meat					26.07	66.22	756.1	10.25	77.5	485.4	562.9	193.2	25.55	64.20
16802	Tail	¼	113.4	1 7	651.0										
	Meat					20.38	72.73	315.2	9.96	31.4	227.0	258.4	56.8	18.02	72.02
	Total	159⅞	72,349.2		72,705.7										

TABLE NO. 2.—*Weights of parts from each cut and data relating to the preparation of air-dry samples.*

PIG No. 1.—BERKSHIRE.

Serial No.	Names of parts and cuts.	Weights of parts.			Preparation of sample.							Removed from original sample.	
		From each cut.	Total.	Of entire pig.	Weight of fresh sample.	Weight of air-dry sample.	Water removed.	Water removed.	Weight of air-dried material taken for extraction.	Weight of extracted sample, air-dry.	Fat removed.	Water removed during extraction.	Weight of original sample equivalent to weight of air-dry sample taken for extraction	Water.	Fat.
		Grams.	*Grams.*	*Per cent.*	*Grams.*	*Grams.*	*Grams.*	*Per cent.*	*Grams.*	*Grams.*	*Grams.*	*Grams.*	*Grams.*	*Per ct.*	*Per ct.*
	Meat (fat and lean):														
	Backs	14,767.9													
	Bellies	8,230.6													
	Hams	9,407.9													
	Shoulders	8,448.2													
	Feet	325.3													
	Spareribs	1,683.8													
	Tenderloins	470.8													
	Neck bones	493.2													
	Backbones	704.0													
	Trimmings	7,021.5													
	Tail	291.7	51,844.9	88.39											
	Bones:														
	Backs	191.1													
	Bellies	81.4													
	Hams	879.6													
	Shoulders	683.8													
	Feet	802.0													
	Spareribs	528.2													
	Neck bones	336.1													
	Backbones	833.5													
	Trimmings	71.0													
	Tail	2.1													
	Total		4,444.4												
16691	Marrow	69.7	69.7	0.12	69.7		9.8							14.06	81.50
16690	Total bones less marrow.		4,374.7	7.44	1,798.0	1,178.0	620.0	34.48	500.0	{ 4.44% 3.1 52.67% 401.8 }	56.8	87.0	11.2	763.1	35.93	11.40

TABLE NO. 2.—*Weights of parts from each cut and data relating to the preparation of air-dry samples*—Continued.

PIG No. 2.—TAMWORTH.

Serial No.	Names of parts and cuts	Weights of parts.			Preparation of sample.									
		From each cut.	Total.	Of entire pig.	Weight of fresh sample.	Weight of air-dry sample.	Water removed.	Water removed.	Weight of air-dried material taken for extraction.	Weight of extracted sample, air-dry.	Fat removed.	Water removed during extraction.	Weight of original sample equivalent to weight of air-dry sample taken for extraction.	Removed from original sample.	
														Water.	Fat.
		Grams.	*Grams.*	*Per cent.*	*Grams.*	*Grams.*	*Grams.*	*Per cent.*	*Grams.*	*Grams.*	*Grams.*	*Grams.*	*Grams.*	*Per ct.*	*Per ct.*
	Meat (fat and lean):														
	Backs	17,201.7													
	Bellies	8,281.8													
	Hams	10,110.2													
	Shoulders	8,256.6													
	Feet	452.3													
	Spareribs	1,575.7													
	Tenderloins	528.2													
	Neck bones	450.5													
	Backbones	860.2													
	Trimmings	6,979.4													
	Tail	599.5	55,305.1	86.50											
	Bones:														
	Backs	222.1													
	Bellies	116.8													
	Hams	1,111.0													
	Shoulders	872.8													
	Feet	1,056.7													
	Spareribs	557.0													
	Neck bones	411.5													
	Backbones	934.0													
	Trimmings	46.0													
	Tail	37.8													
	Total		5,365.7												
16720	Marrow	133.9	133.9	0.21	133.9										
16719	Total bones less marrow.	5,056.0	5,231.8	8.18	2,582.0	1,699.0	883.0	34.20	500.0	2.39% 3.2 50.71% 385.4	113.1 112.6	37.6 -2.0	759.9	13.14 34.46	84.47 14.83

PIG No. 3.—CHESTER WHITE.

Meat (fat and lean):											
Backs	15,217.9										
Bellies	8,966.6										
Hams	7,765.3										
Shoulders	8,468.4										
Feet	269.1										
Spareribs	1,012.8										
Tenderloins	453.6										
Neck bones	427.1										
Backbones	584.5										
Trimmings	6,398.9										
Tail	633.9	50,198.1	£7.94								
Bones:											
Backs	135.8										
Bellies	74.9										
Hams	751.0										
Shoulders	661.1										
Feet	679.0										
Spareribs	396.2										
Neck bones	249.7										
Backbones	556.0										
Trimmings	49.1										
Tail	36.0										
Total		3,588.8									
16631 Marrow	43.6	43.6	0.08	1,740.4	43.6	659.0	37.86	500.0	(4.64%) 2.02	34.82	6.76	15.50	79.86
16630 Total bones less marrow		3,545.2	6.21	1,081.4					(45.56%) 366.6	136.5	−3.1	804.63	37.47	16.97

24

TABLE No. 2.—*Weight of parts from each cut and data relating to the preparation of air-dry samples*—Continued.

PIG No. 4.—POLAND CHINA.

Serial No.	Names of parts and cuts.	Weights of parts.			Preparation of sample.							Removed from original sample.	
		From each cut.	Total.	Of entire pig.	Weight of fresh sample.	Weight of air-dry sample.	Water removed.	Water removed.	Weight of air-dried material taken for extraction.	Weight of extracted sample air-dry.	Fat removed.	Water removed during extraction.	Weight of original sample equivalent to weight of air-dry sample taken for extraction.	Water.	Fat.
		Grams.	*Grams.*	*Per cent.*	*Grams.*	*Grams.*	*Grams.*	*Per cent.*	*Grams.*	*Grams.*	*Grams.*	*Grams.*	*Grams.*	*Per ct.*	*Per ct.*
	Meat (fat and lean):														
	Backs	16,830.2													
	Bellies	10,151.8													
	Hams	10,636.0													
	Shoulders	9,735.5													
	Feet	466.6													
	Spareribs	1,522.4													
	Tenderloins	419.8													
	Neck bones	537.0													
	Backbones	638.6													
	Trimmings	8,574.3													
	Tail	671.6	60,182.7	90.67											
	Bones:														
	Backs	114.3													
	Bellies	56.7													
	Hams	731.1													
	Shoulders	602.1													
	Feet	673.5													
	Spareribs	447.1													
	Neck bones	266.5													
	Backbones	639.0													
	Trimmings	32.7													
	Tail	28.0													
	Total		3,591.0												
16665	Marrow	72.0	72.0	0.11	72.0										
16663	Total bones less marrow.		3,519.0	5.30	1,729.7	1,042.2	687.5	39.75	500.0	(5.28%) 3.8 (50.37%) 418.0	56.4 78.0	11.8 4.0	829.88	16.39 40.23	78.33 9.40

PIG No. 5.—DUROC JERSEY.

Meat (fat and lean):										
Backs	16,709.2									
Bellies	10,189.1									
Hams	8,396.1									
Shoulders	7,971.0									
Feet	200.1									
Spareribs	1,195.4									
Tenderloins	348.5									
Neck bones	488.5									
Backbones	702.3									
Trimmings	8,155.7									
Tail	596.6									
	54,952.5	86.03								
Bones:										
Backs	145.0									
Bellies	55.0									
Hams	698.0									
Shoulders	597.8									
Feet	774.0									
Spareribs	416.6									
Neck bones	283.2									
Backbones	722.3									
Trimmings	43.4									
Tail	30.0									
Total	3,764.5									
16601 Marrow	67.2	0.11	67.2						12.80	80.95
16600 Total bones less marrow	3,697.3	5.92	1,655.0	1,084.8	570.2	34.45	500.0	(6.25%) 4.2 (53.64%) 409.2	54.4 88.3	8.6 2.5	762.78	34.78	11.58

TABLE No. 2.—*Weights of parts from each cut and data relating to the preparation of air-dry samples*—Continued.

PIG No. 6.—DUROC JERSEY.

Serial No.	Names of parts and cuts.	Weights of parts.			Preparation of sample.							Removed from original sample.	
		From each cut.	Total.	Of entire pig.	Weight of fresh sample.	Weight of air-dry sample.	Water removed.	Water removed.	Weight of air-dried material taken for extraction.	Weight of extracted sample, air-dry.	Fat removed.	Water removed during extraction.	Weight of original sample equivalent to weight of air-dry sample taken for extraction.	Water.	Fat.
		Grams.	*Grams.*	*Per cent.*	*Grams.*	*Grams.*	*Grams.*	*Per cent.*	*Grams.*	*Grams.*	*Grams.*	*Grams.*	*Grams.*	*Per ct.*	*Per ct.*
	Meat (fat and lean):														
	Backs	19,194.9													
	Bellies	14,101.6													
	Hams	11,120.6													
	Shoulders	8,171.7													
	Feet	493.7													
	Spareribs	1,214.3													
	Tenderloins	421.3													
	Neck bones	635.6													
	Backbones	801.7													
	Trimmings	10,670.8													
	Tail	1,070.7	68,913.7	90.93											
	Bones:														
	Backs	169.8													
	Bellies	68.5													
	Hams	720.9													
	Shoulders	585.0													
	Feet	671.0													
	Spareribs	397.9													
	Neck bones	242.3													
	Backbones	690.0													
	Trimmings	65.3													
	Tail	27.0													
16751	Total		3,643.7		1,793.5	1,227.5	566.0	31.56	500.0	(51.74%) 378.0	127.7	−5.7	730.57	30.78	17.48
16748	Marrow	79.0	79.0	0.10											
	Total bones less marrow.		3,564.7	4.70											

PIG No. 7.—DUROC JERSEY.

Meat (fat and lean):									
Backs	16,807.7								
Bellies	12,404.8								
Hams	9,746.2								
Shoulders	8,043.4								
Feet	290.8								
Spareribs	1,114.1								
Tenderloins	343.3								
Neck bones	611.4								
Backbones	785.8								
Trimmings	10,029.2								
Tail	672.8	60,889.5	86.90						
Bones:									
Backs	105.0								
Bellies	58.5								
Hams	604.7								
Shoulders	587.0								
Feet	665.2								
Spareribs	389.9								
Neck bones	281.1								
Backbones	654.0								
Trimmings	53.0								
Tail	23.7								
16776 Total		3,502.1	0.11	1,859.3	1,231.8	627.5	33.75	500.0	(53.50%) 403.8	97.2	−1.0	754.72	33.62
Marrow	76.1	76.1		76.1									
16775 Total bones less marrow		3,426.0	5.07										12.88

TABLE No. 2.—*Weight of parts from each cut and data relating to the preparation of air-dry samples*—Continued.

PIG No. 8.—YORKSHIRE.

Serial No.	Names of parts and cuts.	Weights of parts.			Preparation of sample.								
		From each cut.	Total.	Of entire pig.	Weight of fresh sample.	Weight of air-dry sample.	Water removed.	Water removed.	Weight of air-dried material taken for extraction.	Weight of extracted sample air-dry.	Fat removed.	Water removed during extraction.	Weight of original sample equivalent to weight of air-dry sample taken for extraction.	Removed from original sample.	
														Water.	Fat.
		Grams.	*Grams.*	*Per cent.*	*Grams.*	*Grams.*	*Grams.*	*Per cent.*	*Grams.*	*Grams.*	*Grams.*	*Grams.*	*Grams.*	*Per ct.*	*Per ct.*
	Meat (fat and lean):														
	Backs............	18,458.1													
	Bellies...........	9,562.2													
	Hams............	10,682.4													
	Shoulders........	11,124.8													
	Feet.............	673.3													
	Spareribs........	1,774.0													
	Tenderloins......	632.5													
	Neck bones......	770.9													
	Backbones.......	1,100.0													
	Trimmings.......	7,743.6													
	Tail.............	565.5													
			63,087.3	86.79											
	Bones:														
	Backs............	246.3													
	Bellies...........	61.8													
	Hams............	1,220.0													
	Shoulders........	940.2													
	Feet.............	1,089.0													
	Spareribs........	566.0													
	Neck bones......	408.0													
	Backbones.......	898.0													
	Trimmings.......	81.2													
	Tail.............	23.0													
	Total........		5,486.9	0.13											
16905	Marrow..........	92.9	92.9	7.41	92.9										
16904	Total bones less marrow.		5,394.0		2,470.9	1,615.7	855.2	34.61	500.0	{(4.41%) 4.1 (50.51%) 386.2}	75.78 105.1	13.02 8.7	764.64	14.02 35.75	81.57 13.74

TABLE No. 3.—*Weight of parts from each cut and data relating to the preparation of air-dry samples.*

PIG No. 1.—BERKSHIRE.

Serial No.	Names of parts and cuts.	Weight of parts.		Of entire pig.	Direct determinations on original material.		Preparation of samples.						Removed in preparation of sample.	
		From each cut.	Total.		Water.	Fat.	Weight of fresh sample.	Weight of air-dry sample after extraction.	Weight of fat.	Air-dry sample + fat.	Weight of water removed.	Air-dry sample in the original material.	Water.	Fat.
		Grams.	*Grams.*	*Per cent.*	*Per ct.*	*Per ct.*	*Grams.*	*Grams.*	*Grams.*	*Grams.*	*Grams.*	*Per ct.*	*Per ct.*	*Per ct.*
16688	Skin				57.66	16.84	183.6	67.8	29.2	97.0	86.6	36.93	47.17	15.90
	Backs	633.5												
	Bellies	419.8												
	Hams	287.1												
	Shoulders	253.5												
	Foot	174.1												
	Trimmings	420.3												
	Tail	44.2												
			2,232.5	3.80										
16693	Spinal cord	13.2					55.7	4.9	14.5	19.4	36.3	8.80	65.17	26.03
	Neck bones	42.5												
	Backbones		55.7	0.09										
16695	Tendons	159.5	159.5	0.27			191.3	61.1	24.7	85.8	105.5	31.93	55.16	12.91
17177	Hoofs (?): Feet													
	Feet (corrected)	52.6	52.6	0.09			52.6	a33.37			19.23	63.44	36.56	

TABLE No. 3.—*Weight of parts from each cut and data relating to the preparation of air-dry samples*—Continued.

PIG No. 2.—TAMWORTH.

Serial No.	Names of parts and cuts.	Weight of parts.		Of entire pig.	Direct determinations on original material.		Preparation of samples.						
		From each cut.	Total.		Water.	Fat.	Weight of fresh sample.	Weight of air-dry sample after extraction.	Weight of fat.	Air-dry sample + fat.	Weight of water removed.	Air-dry sample in the original material.	Removed in preparation of sample.	
													Water.	Fat.
		Grams.	*Grams.*	*Per cent.*	*Per ct.*	*Per ct.*	*Grams.*	*Grams.*	*Grams.*	*Grams.*	*Grams.*	*Per ct.*	*Per ct.*	*Per ct.*
16717	Skin	947.0			48.74	20.02	240.4	82.4	30.7	113.1	127.3	34.28	52.95	12.77
	Backs	475.0												
	Bellies	459.0												
	Hams	282.8												
	Shoulders	261.9												
	Feet	515.7												
	Trimmings	70.2												
	Tail		3,011.6	4.71										
16722	Spinal cord	15.0												
	Neck bones	45.8					60.8	5.4	27.5	32.9	27.9	8.88	45.88	45.24
	Backbones		60.8	0.09										
16724	Tendons	138.7	138.7	0.21			176.7	59.5	13.8	73.3	103.4	33.67	58.52	7.81
17178	Hoofs (6):													
	Feet (corrected)	64.5	64.5	0.10			64.5	a 39.17			25.33	69.73	39.27	

PIG No. 3.—CHESTER WHITE.

16633	Skin	862.5			40.78	31.17	431.5		b 78.1					a 18.10
	Backs	484.1												
	Bellies	541.5												
	Hams	311.0												
	Shoulders	184.6												
	Feet	696.2												
	Trimmings	70.3												
	Tail		3,150.2	5.59										

PIG No. 4.—POLAND CHINA.

16635	Spinal cord: d								
	Neck bones								
	Backbones	6.9							
		32.1	39.0	0.07					
16637	Tendons	67.2	67.2	0.12	75.1	25.7	7.6	33.3	41.8	34.22	55.66	10.12
17179	Feet									
	Hoofs (4):									
	Feet (corrected)	37.0	37.0	0.06	37.0	a24.19			12.81	65.38	34.62
16659	Skin	632.5								
	Backs	507.6								
	Bellies	369.8								
	Hams	293.5								
	Shoulders	78.0								
	Feet	465.0								
	Trimmings	60.4								
	Tail	2,406.8	38.10	3.63	311.3	104.0	68.8	172.8	138.5	33.41	44.49	22.10
16661	Spinal cord	11.2								
	Neck bones	39.9								
	Backbones		51.1	0.08	50.6	6.2	21.1	27.3	23.3	12.25	46.05	41.70
16664	Tendons	96.0	96.0	0.14	121.5	42.8	13.9	56.7	64.8	35.22	53.34	11.44
17180	Hoofs (8):									
	Feet	44.9	44.9	0.07	44.9	a25.39			19.51	56.55	43.45

a Not extracted.   b Ether extract lost (poured).   c Poured.   d Sample lost.

32

TABLE No. 3.—*Weight of parts from each cut and data relating to the preparation of air-dry samples*—Continued.

PIG No. 5.—DUROC JERSEY.

Serial No.	Names of parts and cuts.	Weight of parts.		Of entire pig.	Direct determinations on original material.		Preparation of samples.						
		From each cut.	Total.		Water.	Fat.	Weight of fresh sample.	Weight of air-dry sample after extraction.	Weight of fat.	Air-dry sample + fat.	Weight of water removed.	Air-dry sample in the original material.	Removed in preparation of sample.	
													Water.	Fat.
		Grams.	*Grams.*	*Per cent.*	*Per ct.*	*Per ct.*	*Grams.*	*Grams.*	*Grams.* (a)	*Grams.*	*Grams.*	*Grams.*	*Per ct.*	*Per ct.*
16603	Skin	1,063.0			35.49	38.16								
	Hacks	699.0												
	Bellies	431.5												
	Hams	419.0												
	Shoulders	190.1												
	Feet	731.2												
	Trimmings	56.4												
	Tail		3,590.2	5.75										
16605	Spinal cord	12.3												
	Neck bones	13.4					23.8	4.3	5.62	9.92	13.88	18.07	58.32	23.61
	Backbones		25.7	0.04										
16607	Tendons:b													
	Feet	59.6	59.6	0.10										
17181	Hoofs (7):													
	Feet (corrected)	31.6	31.6	0.05			31.6	c 22.13			9.47	70.03	29.97	

PIG No. 6.—DUROC JERSEY.

16746	Skin	820.5			45.20	20.59	420.2	152.5	d 28.1					
	Hacks	657.0												
	Bellies	340.0												
	Hams	279.2												
	Shoulders	230.8												
	Feet	623.5												
	Trimmings	75.3												
	Tail		3,035.3	4.00										
16749	Spinal cord: b													
	Neck bones	14.7												
	Backbones	40.3	55.0	0.07										
16753	Tendons	92.5	92.5	0.12			92.5	31.92	8.56	40.48	52.02	34.51	56.24	9.25
17182	Hoofs (8):													
	Feet	59.2	59.2	0.08			59.2	c 34.83			24.37	58.83	41.17	

PIG No. 7.—DUROC JERSEY.

16778	Skin			49.26	16.39	447.0	175.2	63.8	239.0	208.0	39.18	46.55	14.27
	Backs	721.0										
	Bellies	691.1										
	Hams	370.5										
	Shoulders	328.2										
	Feet	314.0										
	Trimmings	662.5										
	Tail	62.5	3,169.8	4.65								
16780	Spinal cord	14.0										
	Neck bones	42.2				56.2	7.17	37.78	44.95	11.25	12.76	20.02	67.22
	Backbones		56.2	0.08								
16782	Tendons	78.0	78.0	0.11		78.0	24.91	10.82	35.73	42.27	31.94	54.19	13.87
	Feet											
17183	Hoofs (1): Feet (corrected)	52.0	52.0	0.08		52.0	c31.23			20.77	60.06	39.94

PIG No. 8.—YORKSHIRE.

16807	Skin			52.50	11.71	(b)						
	Backs	1,169.0										
	Bellies	837.2										
	Hams	458.2										
	Shoulders	431.4										
	Feet	275.2										
	Trimmings	623.5										
	Tail	62.5	3,857.0	5.30								
16809	Spinal cord: b	13.4										
	Neck bones	52.6										
	Backbones		66.0	0.09								
16811	Tendons: b Feet	133.5	133.5	0.18								
17184	Hoofs (3): Feet	75.0	75.0	0.10		75.0	c38.27			36.73	51.03	48.97

a Ether extract lost.   *b* Sample lost.   *c* Not extracted.   *d* Ether extract lost (poured).

TABLE NO. 4.—Ana

PIG No. 1.—BERKSHIRE.

Serial No.	Names of cuts.	Air-dry sample: Per cent of original material.	Per cent air-dry material.						
			Water.	Fat.	Total.	Nitrogen.			
						Of proteids insoluble in hot water.	Precipitated by bromin.	Of flesh bases.	Lecithin.	Ash.
16667	2 American backs	13.16	3.14	20.55	11.32	8.51	0.62	2.19	1.16	3.89
16669	2 clear bellies	14.33	3.14	21.59	11.15	7.78	0.65	2.72	0.99	3.85
16671	2 short-cut hams	22.95	4.14	15.43	11.85	9.77	0.48	1.60	1.10	4.18
17165	(Fat extracted with ether)	16.58			0.22				2.43	
16673	2 New York shoulders	17.05	2.31	2.10	13.76	10.22	0.73	2.81	0.85	5.03
16675	4 feet	25.10	6.46	6.32	13.73	7.75	3.00	2.98	0.75	3.28
17174	(Fat extracted with ether)	15.20			0.13				2.68	
16677	Spareribs	20.81	3.66	8.23	13.03	10.31	0.89	1.83	1.68	4.80
16679	Tenderloins	27.11	5.14	9.47	12.50	10.95	0.28	1.27	1.82	4.30
16680	Neck bones	20.02	7.23	10.93	12.25	9.97	0.59	1.69	1.33	4.02
17159	(Fat extracted with ether)	18.69			0.21				2.17	
16682	Backbones	22.24	3.36	6.88	13.03	10.36	0.62	2.05	1.20	5.59
16684	Trimmings	9.72	3.69	8.34	13.09	8.54	1.11	3.44	1.16	4.23
16686	Tail	8.73	4.30	6.97	13.45	10.56	0.98	1.91	1.98	4.41

PIG No. 2.—TAMWORTH.

16696	2 American clear backs	10.48	3.46	12.22	12.61	8.74	1.08	2.79	1.23	4.06
16698	2 American clear bellies	12.12	4.27	14.91	12.44	9.19	0.86	2.39	1.23	3.90
16700	2 short-cut hams	19.99	5.38	8.79	12.92	10.41	0.50	2.01	1.11	4.22
16702	2 New York shoulders	20.36	3.54	19.93	11.44	8.86	0.70	1.88	1.20	3.86
16704	4 feet	23.85	6.24	4.59	14.04	7.81	2.25	3.98	1.57	3.60
17149	(Fat extracted with ether)	15.50			0.16				0.87	
16706	Spareribs	19.67	5.49	8.51	12.75	9.38	1.06	2.31	1.27	4.75
16708	Tenderloins	24.12	4.74	8.38	12.92	11.37	0.37	1.18	1.46	4.40
17131	(Fat extracted with ether)	11.49			0.22				2.98	
16709	Neck bones	20.95	5.44	6.49	13.36	10.24	0.98	2.14	1.24	4.88
17164	(Fat extracted with ether)	19.43			0.14				2.91	

lytical data for meats.

PIG No. 1.—BERKSHIRE.

Per cent original material.

Water.			Fat.			Lecithin.	Total.	Nitrogen.				Nitrogenous substances.				Ash.	Total.[1]
In preparing sample.	In air-dry material.	Total.	In preparing sample.	In air-dry material.	Total.			Of proteids insoluble in hot water.	Precipitated by bromin.	Of flesh bases.		Proteids insoluble in hot water.	Gelatinoids.	Flesh bases.	Total.		
31.86	0.41	32.27	54.98	2.71	57.09	0.15	1.49	1.12	0.08	0.29		7.00	0.50	0.91	8.41	0.51	98.46 / 98.88
36.82	0.45	37.27	48.84	3.09	51.93	0.14	1.60	1.12	0.09	0.39		7.00	0.50	1.22	8.78	0.55	98.11 / 98.53
60.47	0.95	61.42	16.58	3.54	20.12	0.25	2.72	2.24	0.11	0.37		14.00	0.69	1.15	15.84	0.96	99.28 / 98.34
						0.40	0.04										
53.64	0.41	54.04	28.71	0.37	29.08	0.15	2.43	1.80	0.13	0.50		11.25	0.81	1.50	13.62	0.89	98.49 / 97.63
59.66	1.62	61.28	15.24	1.59	16.83	0.20	3.45	1.95	0.75	0.75		12.19	4.69	2.34	19.22	0.82	96.86 / 98.15
						0.41	0.02										
51.78	0.76	52.54	27.39	1.71	29.10	0.35	2.71	2.15	0.18	0.38		13.44	1.13	1.19	15.76	1.00	97.14 / 98.40
66.67	1.39	68.06	6.21	2.57	8.78	0.40	3.30	2.97	0.08	0.34		18.56	0.50	1.06	20.12	1.17	97.57 / 98.13
54.25	1.45	55.70	25.73	2.19	27.92	0.27	2.45	1.99	0.12	0.34		12.44	0.75	1.06	14.25	0.81	97.60 / 98.68
						0.41	0.04										
52.08	0.75	52.83	25.09	1.53	27.22	0.20	2.90	2.30	0.14	0.46		14.38	0.87	1.44	16.69	1.24	96.98 / 97.98
29.11	0.36	29.47	61.17	0.81	61.98	0.11	1.27	0.83	0.11	0.33		5.19	0.69	1.03	6.91	0.41	99.00 / 98.77
23.64	0.38	24.02	67.62	0.61	68.23	0.17	1.17	0.92	0.09	0.16		5.75	0.56	0.50	6.81	0.39	100.44 / 99.45

PIG No. 2.—TAMWORTH.

41.48	0.36	41.84	48.04	1.28	49.32	0.13	1.32	0.92	0.11	0.29		5.75	0.60	0.01	7.35	0.43	98.67 / 98.94
33.17	0.52	33.69	54.71	1.81	56.52	0.15	1.51	1.12	0.10	0.29		7.00	0.63	0.91	8.54	0.47	100.81 / 99.22
		1.08			1.76	0.22	2.58	2.08	0.10	0.40		13.00	0.63	1.25	14.88	0.84	98.10
		0.72			4.06	0.24	2.33	1.80	0.14	0.38		11.25	0.87	1.19	13.31	0.79	79.15
58.39	1.49	59.88	17.76	1.09	18.85	0.37	3.35	1.86	0.54	0.95		11.63	3.38	2.96	17.97	0.86	98.72 / 97.56
						0.14	0.03										
48.12	1.08	49.20	32.21	1.67	33.88	0.51	2.51	1.85	0.21	0.45		11.56	1.31	1.40	14.27	0.93	84.57 / 98.28
64.38	1.14	65.52	11.49	2.02	13.51	0.25	3.12	2.74	0.09	0.29		17.13	0.56	0.91	18.60	1.06	97.27 / 98.09
						0.35											
						0.56	0.03										
54.38	1.14	55.52	24.67	1.36	26.03	0.91	2.80	2.14	0.21	0.45		13.38	1.31	1.40	16.09	1.02	98.46 / 98.66
						0.26											
						0.57	0.03										
						0.83											

[1] In this column the totals obtained by both the direct and the indirect determination of water and fat are given. The upper number in each case was obtained by use of the results of direct determinations of these constituents; for the lower number in each case the results obtained during the preparation of the sample, and in the analysis of the dry-air sample, were used. Lecithin is not included in the totals given in this table.

TABLE No. 4.—*Ana*

PIG No. 2.—TAMWORTH—Continued.

Serial No.	Names of cuts.	Air-dry sample: Per cent of original material.	Per cent air-dry material.							
			Water.	Fat.	Total.	Nitrogen.		Of flesh bases.	Lecithin.	Ash.
						Of proteids insoluble in hot water.	Precipitated by bromin.			
16711	Backbones	20.98	6.50	6.36	12.95	10.44	0.68	1.83	1.19	5.25
16713	Trimmings	10.53	4.83	14.61	12.41	8.08	1.24	3.09	0.98	4.05
16715	Tail	9.53	4.83	20.15	10.90	7.13	1.15	2.62	1.09	3.15

PIG No. 3.—CHESTER WHITE.

16609	2 American clear backs	10.09	2.18	22.71	10.17	7.12	0.73	2.32	1.23	3.47
16611	2 American clear bellies	9.81	2.23	8.75	13.20	8.82	1.01	3.37	0.85	4.31
16613	2 short-cut hams	24.30	2.65	32.10	10.34	7.34	0.42	2.58	1.42	3.28
16615	2 New York shoulders	17.75	9.86	12.01	11.69	8.50	0.78	2.41	1.57	4.01
16617	4 feet	20.71	4.00	3.32	14.21	7.64	4.92	1.65	0.93	4.08
16619	Spareribs	26.69	2.95	26.47	10.67	8.17	0.52	1.98	1.04	3.45
16621	Tenderloins	25.88	3.89	16.65	11.88	10.32	0.27	1.20	1.55	4.00
16622	Neck bones	19.30	2.79	9.24	12.89	10.11	0.59	2.19	1.28	4.51
16624	Backbones	19.39	1.88	5.53	13.73	11.05	0.48	2.20	1.19	5.43
16626	Trimmings	7.20	3.87	13.38	11.90	7.16	1.15	3.68	0.99	4.29
16628	Tail	5.65	3.78	9.14	13.29	8.42	1.25	3.62	1.33	4.10

PIG No. 4.—POLAND CHINA.

16638	2 American clear backs	9.68	3.21	16.05	12.10	8.50	0.93	2.67		3.89
16640	2 American clear bellies	9.43	4.26	5.34	13.23	8.14	1.09	4.00	1.31	4.53
16642	2 short-cut hams	17.00	4.79	6.38	13.23	10.07	0.75	2.41	1.33	4.48
16644	2 New York shoulders		15.28	8.00	11.34	8.62	0.68	2.04	1.45	4.07
17152	(Fat extracted with ether) *a*				0.09				2.03	
16646	4 feet	19.57	7.29	0.50	14.01	9.02	3.42	1.57	0.77	4.64
16648	Spareribs	23.03	4.83	16.65	11.71	6.99	0.42	4.30	1.33	4.11
16650	Tenderloins	26.80	5.88	13.40	11.85	10.53	0.41	0.91	1.45	4.22
16651	Neck bones	19.37	5.96	8.89	12.27	8.46	0.93	2.88	0.89	4.71
17163	(Fat extracted with ether)	20.67			0.24				1.36	
16653	Backbones	21.21	3.68	12.07	12.42	9.77	0.62	2.03	1.21	4.95
16655	Trimmings	8.05	3.54	21.29	10.98	7.39	0.87	2.72	0.80	3.01
16657	Tail	6.41	5.48	5.21	13.23	8.61	1.32	3.30	1.19	4.90

a Fat extract lost.

lytical data for meats—Continued.

PIG No. 2.—TAMWORTH—Continued.

Per cent original material.															
Water.			Fat.			Lecithin.	Total.	Nitrogen.				Nitrogenous substances.			
In preparing sample.	In air-dry material.	Total.	In preparing sample.	In air-dry material.	Total.			Of proteids insoluble in hot water.	Precipitated by bromin.	Of flesh bases.	Proteids insoluble in hot water.	Gelatinoids.	Flesh bases.	Total.	Ash.	Total.
49.68	1.38	51.06	29.34	1.33	30.67	0.25	2.72	2.19	0.14	0.39	13.69	0.87	1.22	15.78	1.10	97.66 / 98.61
28.34	0.51	28.85	61.13	1.54	62.67	0.10	1.31	0.85	0.13	0.33	5.31	0.81	1.03	7.15	0.43	101.30 / 99.10
25.31	0.46	25.77	65.16	1.92	67.08	0.10	1.04	0.68	0.11	0.25	4.25	0.69	0.78	5.72	0.30	97.14 / 98.87

PIG No. 3.—CHESTER WHITE.

71.88	0.22	72.10	18.02	2.29	20.31	0.12	1.03	0.72	0.07	0.24	4.50	0.44	0.75	5.69	0.35	99.92 / 98.45
30.32	0.22	30.54	59.87	0.86	60.73	0.08	1.30	0.87	0.10	0.33	5.44	0.63	1.03	7.10	0.42	99.15 / 98.79
52.51	0.64	53.15	23.19	7.80	30.99	0.35	2.51	1.78	0.10	0.03	11.13	0.63	1.97	13.73	0.80	97.93 / 98.67
	1.75			2.13		0.28	2.08	1.51	0.14	0.43	9.44	0.87	1.34	11.65	0.71	99.14
	0.83			0.69		0.19	2.94	1.58	1.02	0.34	9.88	6.38	1.06	17.32	0.84	97.95
52.44	0.79	53.23	20.87	7.06	27.93	0.28	2.85	2.18	0.14	0.53	13.63	0.87	1.65	16.15	0.92	97.67 / 98.23
64.96	1.01	65.97	9.16	4.31	13.47	0.40	3.07	2.67	0.07	0.33	16.69	0.44	1.03	18.16	1.06	97.94 / 98.60
52.96	0.54	53.50	27.74	1.78	29.52	0.25	2.49	1.95	0.12	0.42	12.19	0.75	1.31	14.25	0.87	96.35 / 98.14
50.03	0.36	50.39	30.58	1.07	31.65	0.23	2.66	2.14	0.09	0.43	13.38	0.56	1.34	15.28	1.05	98.18 / 98.37
22.21	0.28	22.49	70.59	0.96	71.55	0.07	0.87	0.52	0.08	0.27	3.25	0.50	0.84	4.59	0.31	100.11 / 98.94
	0.21			0.52		0.08	0.75	0.48	0.07	0.20	3.00	0.44	0.62	4.06	0.24	99.65

PIG No. 4.—POLAND CHINA.

29.22	0.31	29.53	61.10	1.55	62.65		1.17	0.82	0.09	0.26	5.13	0.56	0.81	6.50	0.38	99.34 / 99.06
30.38	0.40	30.78	60.19	0.50	60.69	0.12	1.25	0.77	0.10	0.38	4.81	0.63	1.19	6.63	0.43	98.93 / 98.53
53.97	0.81	54.78	29.03	1.09	30.12	0.23	2.25	1.71	0.13	0.41	10.69	0.81	1.28	12.78	0.76	98.00 / 98.44
		51.72			33.74	0.30	1.65	1.25	0.10	0.30	7.81	0.63	0.94	9.38	0.59	95.43
49.23	1.43	50.66	31.20	0.12	31.32	0.15	2.74	1.76	0.67	0.31	11.00	4.19	0.97	16.16	0.91	98.36 / 99.05
55.53	1.11	56.64	22.44	3.83	26.27	0.31	2.70	1.61	0.10	0.99	10.06	0.63	3.00	13.78	0.95	97.23 / 97.64
65.85	1.58	67.43	7.36	3.59	10.95	0.39	3.18	2.82	0.11	0.25	17.63	0.69	0.78	19.10	1.13	98.69 / 98.61
54.08	1.15	55.23	26.55	1.72	28.27	0.17	2.38	1.64	0.18	0.56	10.25	1.13	1.75	13.13	0.91	97.86 / 97.54
						0.28	0.05									
						0.45										
48.65	0.78	49.43	30.14	2.56	32.70	0.27	2.63	2.07	0.13	0.43	12.94	0.81	1.34	15.09	1.05	98.38 / 98.27
22.14	0.29	22.43	69.81	1.71	71.52	0.06	0.88	0.59	0.07	0.22	3.69	0.44	0.69	4.82	0.32	99.29 / 99.09
16.15	0.35	16.50	77.44	0.33	77.77	0.08	0.85	0.55	0.09	0.21	3.44	0.56	0.66	4.66	0.31	100.36 / 99.24

TABLE NO. 4.—*Ana*

PIG No. 5.—DUROC JERSEY.

Serial No.	Names of cuts.	Air-dry sample: Per cent of original material.	Per cent air-dry material.							
			Water.	Fat.	Total.	Nitrogen.			Of flesh bases.	Lecithin.	Ash.
						Of proteids insoluble in hot water.	Precipitated by bromin.			
16579	2 American clear backs	9.37	9.69	24.48	10.47	6.76	0.81	2.90	0.82	3.14
16581	2 clear bellies	11.71	3.05	28.08	10.39	6.84	0.80	2.75	0.83	3.57
16583	2 short-cut hams	21.39	3.48	32.03	7.75	6.25	0.73	0.77	0.16	3.34
16585	2 New York shoulders		3.08	4.39	13.82	10.11	0.83	2.88	0.84	4.57
16587	4 feet	23.36	5.21	7.88	12.47	6.97	1.52	3.98		3.27
16589	Spareribs	22.23	3.21	11.30	12.92	10.14	0.56	2.22	0.53	4.67
16591	Tenderloins	28.54	4.23	18.53	11.69	10.25	0.22	1.22	1.22	3.98
16592	Neck bones	19.46	3.93	7.58	13.20	10.28	0.61	2.31	1.05	4.85
16594	Backbones	25.94	3.44	12.41	12.64	10.08	0.50	2.06	1.52	4.75
16596	Trimmings	6.80	3.98	10.90	12.77	7.66	0.95	4.16	1.14	4.38
16598	Tail	4.70	4.37	13.88	12.19	6.85	1.10	4.24	1.39	4.30

PIG No. 6.—DUROC JERSEY.

16725	2 American clear backs	6.99	6.75	6.71	12.89	8.50	1.07	3.23	0.97	4.17
17133	(Fat extracted with ether)	20.65			0.13				0.59	
16727	2 clear bellies	8.30	4.96	16.65	11.79	8.44	1.04	2.31	0.97	2.59
17134	(Fat extracted with ether)	6.17			0.09				0.51	
16729	2 short cut hams	27.63	5.49	8.95	12.75	10.05	0.73	1.97	1.28	4.76
16731	2 New York shoulders	12.91	6.58	5.39	13.09	9.30	1.12	2.67	0.92	6.53
16733	4 feet	21.84	7.68	6.72	13.23	7.72	2.11	3.40	0.74	3.44
17142	(Fat extracted with ether)	16.13			0.12				1.15	
16735	Spareribs	22.09	5.79	11.79	12.27	9.89	0.80	1.58	1.51	4.96
16737	Tenderloins	26.77	5.67	14.63	11.77	10.35	0.38	1.04	2.09	3.93
16738	Neck bones	17.37	6.02	3.11	13.59	10.77	0.89	1.93	1.61	4.98
16740	Backbones	21.13	4.90	11.96	12.27	9.68	0.82	1.77	1.69	4.35
16742	Trimmings	5.85	3.96	15.48	12.19	8.44	1.07	2.68	1.15	4.18
16744	Tail	6.37	0.12	23.97	10.70	7.53	1.11	2.06	0.92	3.68

lytical data for meats—Continued.

PIG No. 5.—DUROC JERSEY.

Water.			Fat.			Lecithin.	Total.	Nitrogen.				Nitrogenous substances.			Total.	Ash.	Total.
In preparing sample.	In air-dry material.	Total.	In preparing sample.	In air-dry material.	Total.			Of proteids insoluble in hot water.	Precipitated by bromin.	Of flesh bases.	Proteids insoluble in hot water.	Gelatinoids.	Flesh bases.				
.....	0.91	.....	.....	2.29	.....	0.08	0.98	0.63	0.08	.27	3.94	0.50	0.84	5.28	0.29	99.57 .....
28.77	0.36	29.13	59.52	3.31	62.83	0.10	1.22	0.81	0.09	0.32	5.06	0.56	1.00	6.62	0.42	100.06 99.00
49.71	0.74	50.45	28.90	7.04	35.94	0.03	1.06	1.34	0.16	0.16	8.38	1.00	0.50	9.88	0.71	95.89 96.98
.....	.....	44.16	.....	.....	43.74	0.10	1.67	1.22	0.10	0.35	7.63	0.63	1.09	9.35	0.55	97.80
53.37	1.22	54.59	23.27	1.84	25.11	.....	2.91	1.63	0.35	0.93	10.19	2.19	2.90	15.28	0.76	96.39 95.74
53.38	0.71	54.09	24.39	2.51	26.90	0.12	2.97	2.25	0.13	0.49	14.06	0.81	1.53	16.40	1.04	98.41 98.43
65.34	1.21	66.55	6.12	5.29	11.41	0.35	3.34	2.93	0.06	0.35	18.31	0.38	1.09	19.78	1.14	98.86 98.88
51.71	0.77	52.48	28.83	1.48	30.31	0.20	2.57	2.00	0.12	0.45	12.50	0.75	1.40	14.65	0.94	98.63 98.38
47.88	0.80	48.77	26.18	3.22	29.40	0.41	3.28	2.62	0.13	0.53	16.38	0.81	1.65	18.84	1.23	102.50 98.24
19.81	0.27	20.08	73.39	0.74	74.13	0.08	0.87	0.52	0.07	0.28	3.25	0.44	0.87	4.56	0.30	98.89 99.07
11.33	0.21	11.54	83.97	0.65	84.62	0.07	0.57	0.32	0.05	0.20	2.00	0.31	0.62	2.03	0.20	99.46 99.29

PIG No. 6.—DUROC JERSEY.

Water.			Fat.			Lecithin.	Total.	Nitrogen.				Nitrogenous substances.			Total.	Ash.	Total.
In preparing sample.	In air-dry material.	Total.	In preparing sample.	In air-dry material.	Total.			Of proteids insoluble in hot water.	Precipitated by bromin.	Of flesh bases.	Proteids insoluble in hot water.	Gelatinoids.	Flesh bases.				
19.85	0.47	20.32	73.16	0.47	73.63	0.08	0.90	0.60	0.07	0.23	3.75	0.44	0.72	4.91	0.29	100.03 99.15
.....	.....	.....	.....	.....	.....	0.12	0.03	.....	.....	.....	.....	.....	.....	.....	.....	.....
34.11	0.41	34.52	57.59	1.38	58.97	0.20 0.08 0.04	0.98 0.01	0.70	0.09	0.19	4.38	0.56	0.59	5.53	0.22	80.36 99.24
35.74	1.52	37.26	36.63	2.47	39.10	0.12 0.35	3.52	2.78	0.20	0.54	17.38	1.25	1.69	20.32	1.32	100.26 98.00
38.61	0.85	39.46	48.48	0.70	49.18	0.12	1.69	1.20	0.15	0.34	7.50	0.94	1.06	9.50	0.84	99.62 98.98
50.16	1.68	51.84	28.00	1.47	29.47	0.16 0.19	2.89 0.02	1.69	0.46	0.74	10.56	2.88	2.31	15.75	0.75	98.14 97.81
48.56	1.28	49.84	29.35	2.60	31.95	0.35 0.33	2.71	2.18	0.18	0.35	13.63	1.13	1.09	15.85	1.10	98.90 98.74
62.34	1.52	63.86	10.89	3.92	14.81	0.56	3.15	2.77	0.10	0.28	17.31	0.63	0.87	18.81	1.05	97.27 98.53
48.25	1.05	49.30	34.38	0.54	34.92	0.28	2.30	1.87	0.15	0.34	11.69	0.94	1.06	13.69	0.87	98.94 98.78
26.09	1.05	27.14	52.78	2.53	55.31	0.36	2.59	2.05	0.17	0.37	12.81	1.06	1.15	15.02	0.92	90.44 98.30
16.28	0.23	16.51	77.87	0.91	78.78	0.07	0.71	0.49	0.06	0.16	3.06	0.38	0.50	3.94	0.25	102.57 99.47
13.93	0.01	13.94	79.70	1.53	81.23	0.06	0.68	0.48	0.07	0.13	3.00	0.44	0.41	3.85	0.23	100.87 99.25

TABLE No. 4.—*Ana*

PIG No. 7.—DUROC JERSEY.

| Serial No. | Names of cuts. | Air-dry sample: Per cent of original material. | Per cent air-dry material. |||||||
| | | | Water. | Fat. | Total. | Nitrogen. ||| Lecithin. | Ash. |
						Of proteids insoluble in hot water.	Precipitated by bromin.	Of flesh bases.		
16754	2 American clear backs	7.39	8.07	12.48	12.24	8.85	1.16	2.23	1.38	3.92
17155	(Fat extracted with ether)	20.82			9.05				1.83	
16756	2 clear bellies		5.29	15.79	12.44	9.12	1.18	2.14	1.52	4.29
17156	(Fat extracted with ether)	21.18			0.06				3.28	
16758	2 short cut hams	14.73	8.23	2.11	13.11	10.39	0.95	1.77	1.25	4.44
17155	(Fat extracted with ether)	24.80			0.14				1.09	
16760	2 New York shoulders	14.90	6.28	8.25	12.58	9.72	1.00	1.86	1.66	6.13
17137	(Fat extracted with ether)	26.49			0.11				1.38	
16762	4 feet	21.46	6.13	6.39	13.48	9.24	1.94	2.30	0.84	3.03
17141	(Fat extracted with ether)	18.12			0.11				1.00	
16764	Spareribs	21.77	5.46	5.94	13.14	10.72	0.80	1.62	1.67	4.64
17154	(Fat extracted with ether)	20.12			0.07				2.35	
16706	Tenderloins	20.76	4.35	9.84	12.58	11.32	0.38	0.88	1.55	4.16
16767	Neck bones	20.09	5.65	9.79	11.74	9.83	0.89	1.02	1.23	4.69
17145	(Fat extracted with ether)	22.72			0.14				1.30	
16769	Backbones	22.45	4.89	11.19	12.33	9.87	0.75	1.71	1.26	4.55
17144	(Fat extracted with ether)	20.33			0.12				0.77	
16771	Trimmings	7.35	5.88	13.23	11.97	7.30	1.42	3.25		3.84
17139	(Fat extracted with ether)	18.10			0.09				0.63	
16773	Tail	5.94	5.00	18.81	11.24	7.65	1.30	2.29	3.02	4.49
17138	(Fat extracted with ether)	27.46			0.09				1.10	

lytical data for meats—Continued.

PIG No. 7.—DUROC JERSEY.

Water.			Fat.			Nitrogen.					Nitrogenous substances.					
In preparing sample.	In air-dry material.	Total.	In preparing sample.	In air-dry material.	Total.	Lecithin.	Total.	Of proteids insoluble in hot water.	Precipitated by bromin.	Of flesh bases.	Proteids insoluble in hot water.	Gelatinoids.	Flesh bases.	Total.	Ash.	Total.
19.50	0.64	20.23	73.03	0.92	73.95	0.10	0.90	0.65	0.09	0.16	4.06	0.56	0.50	5.12	0.29	101.58 / 99.59
						0.38	0.01									
		21.53			73.50	0.48										98.80
						0.32	0.01	0.45	0.06	0.10	2.81	0.38	0.31	3.50	0.21	
						0.16	0.61									
42.87	1.21	44.08	42.40	0.31	42.71	0.48										99.53 / 98.68
						0.18	1.93	1.53	0.14	0.26	9.56	0.87	0.81	11.24	0.65	
						0.27	0.04									
43.68	0.94	44.62	41.42	1.23	42.65	0.45										99.94 / 99.05
						0.25	1.88	1.45	0.15	0.28	9.06	0.94	0.87	10.87	0.91	
						0.37	0.03									
54.66	1.32	55.98	23.88	1.37	25.25	0.02										97.51 / 98.55
						0.16	2.89	1.98	0.42	0.49	12.38	2.63	1.53	16.54	0.78	
						0.18	0.02									
52.01	1.19	53.20	26.22	1.20	27.51	0.36										98.17 / 98.50
						0.36	2.86	2.33	0.18	0.35	14.56	1.13	1.09	16.78	1.01	
						0.47	0.01									
71.93	0.90	72.83	7.31	2.04	9.35	0.83										93.88 / 98.79
						0.32	2.61	2.35	0.08	0.18	14.09	0.50	0.56	15.75	0.86	
50.96	1.14	52.10	28.95	1.97	30.92	0.25	2.36	1.98	0.18	0.20	12.38	1.13	0.62	14.13	0.94	96.01 / 98.09
						0.30	0.03									
51.99	1.10	53.09	25.56	2.51	28.07	0.55										99.61 / 98.31
						0.28	2.77	2.22	0.17	0.38	13.88	1.06	1.19	16.13	1.02	
						0.16	0.02									
20.06	0.43	20.49	72.59	0.97	73.56	0.44	0.88	0.54	0.10	0.24	3.38	0.63	0.75	4.76	0.28	99.75 / 99.09
						0.11	0.010									
13.43	0.30	13.73	80.03	1.12	81.74	0.18	0.67	0.45	0.08	0.14	2.81	0.50	0.44	3.75	0.27	102.28 / 99.49
						0.30	0.025									
						0.48										

TABLE NO. 4.—*Ana*

PIG No. 8.—YORKSHIRE.

Serial No.	Names of cuts.	Air-dry sample: Per cent of original material.	Per cent air-dry material.						
			Water.	Fat.	Total.	Nitrogen.			
						Of proteids insoluble in hot water.	Precipitated by bromin.	Of flesh bases.	Lecithin.	Ash.
16783	2 American clear backs	9.67	5.82	1.95	13.68	9.83	1.74	2.11	1.11	4.28
17160	(Fat extracted with ether)	25.05			0.12				0.84	
16785	2 clear bellies	11.09	6.36	3.54	13.43	9.69	1.85	1.91	1.00	4.54
17150	(Fat extracted with ether)	29.54			0.10				0.45	
16787	2 short cut hams	21.97	5.12	23.74	10.61	8.38	0.62	1.61	1.29	3.37
17140	(Fat extracted with ether)	18.11			0.08				0.75	
16789	2 New York shoulders		8.13	3.67	12.92	9.46	0.98	2.48	1.16	4.37
16791	4 feet		7.04	6.28	13.48	6.99	2.25	4.24	0.67	3.49
16793	Spareribs	22.26	6.22	11.08	12.47	9.76	0.78	1.93	1.47	4.72
16795	Tenderloins	25.23	4.75	14.80	11.96	10.13	0.51	1.32	1.85	4.14
17143	(Fat extracted with ether)	9.33			0.14				2.20	
16796	Neck bones	19.92	6.33	2.50	13.34	10.71	0.79	1.84	1.33	4.98
17135	(Fat extracted with ether)	17.26			0.15				1.64	
16798	Backbones	23.55	5.61	10.58	12.19	9.59	0.84	1.76	1.48	4.99
17157	(Fat extracted with ether)	16.68			0.12				2.60	
16800	Trimmings	10.25	5.57	15.72	11.63	7.10	1.59	2.94	1.20	3.80
17151	(Fat extracted with ether)	30.31			0.09				1.05	
16802	Tail	9.96	4.83	23.55	10.81	7.86	1.68	1.27	1.09	3.33
17147	(Fat extracted with ether)	43.40			0.08				1.02	

lytical data for meats—Continued.

PIG No. 8.—YORKSHIRE.

Water.			Fat.			Per cent original material.					Nitrogenous substances.					
							Nitrogen.									
In preparing sample.	In air-dry material.	Total.	In preparing sample.	In air-dry material.	Total.	Lecithin.	Total.	Of proteids insoluble in hot water.	Precipitated by bromin.	Of flesh bases.	Proteids insoluble in hot water.	Gelatinoids.	Flesh bases.	Total.	Ash.	Total.
27.99	0.50	28.55	62.34	0.19	62.53	0.11	1.32	0.95	0.17	0.20	5.94	1.06	0.62	7.62	0.41	{ 99.91 99.11
						0.21	0.03									
						0.32										
33.08	0.71	33.79	55.83	0.39	56.22	0.12	1.49	1.08	0.20	0.21	6.75	1.25	0.66	8.66	0.48	{ 79.69 90.15
						0.13	0.03									
						0.25										
58.02	1.12	59.14	20.01	5.22	25.23	0.28	2.33	1.84	0.14	0.35	11.50	0.87	1.09	13.46	0.74	{ 97.64 98.57
						0.14	0.015									
						0.42										
		49.57			36.09	0.17	1.85	1.36	0.14	0.35	8.50	0.87	1.09	10.46	0.63	96.75
		57.47			30.86	0.08	1.57	0.82	0.26	0.49	5.13	1.63	1.53	8.29	0.41	97.03
50.93	1.38	52.31	26.81	2.47	29.28	0.33	2.78	2.17	0.18	0.43	13.56	1.13	1.34	16.03	1.05	{ 98.81 98.67
65.45	1.20	66.65	9.32	3.75	13.07	0.47	3.02	2.56	0.13	0.33	16.00	0.81	1.03	17.84	1.04	{ 97.47 98.60
						0.21										
						0.68										
53.87	1.26	55.13	26.21	0.50	26.71	0.26	2.66	2.13	0.16	0.37	13.31	1.00	1.15	15.46	0.99	{ 98.99 98.29
						0.28	0.026									
						0.54										
49.33	1.32	50.65	27.12	2.49	29.61	0.35	2.87	2.26	0.20	0.41	14.13	1.25	1.28	16.66	1.18	{ 98.26 98.10
						0.43	0.02									
						0.78										
25.55	0.57	26.12	64.20	1.61	65.81	0.12	1.19	0.73	0.16	0.30	4.56	1.00	0.94	6.50	0.39	{ 99.18 98.82
						0.32	0.027									
						0.44										
18.02	0.48	18.50	72.02	2.35	74.37	0.11	1.08	0.78	0.17	0.13	4.88	1.06	0.41	6.35	0.33	{ 99.79 99.55
						0.44	0.035									
						0.55										

TABLE No. 5.—*Analytical data for bones,*

PIG No. 1.—BERKSHIRE.

Serial No.	Names of parts.	Air-dry sample, per cent of original material.	Water.	Fat.	Total.	Of proteids insoluble in hot water.	Precipitated by bromin.	Of flesh bases.	Lecithin.	Ash.
						Nitrogen.				
16690	Bones	52.67	5.72	0.52	6.18	5.32	0.11	0.75	0.84	49.59
16690A	(Fat extracted with ether)	11.40	(0.61)	.....	0.29	.....	.....	.....	.....	(0.28)
16691	Marrow	4.44	6.68	0.19	8.31	7.08	0.65	0.48	.....	.....
17169	(Fat extracted with ether)	17.30	.....	.....	0.07	.....	.....	.....	2.64	.....
16688	Skin	36.93	8.31	3.28	15.02	10.95	2.89	1.18	0.33	1.70
17175	(Fat extracted with ether)	15.90	.....	.....	0.15	.....	.....	.....	1.85	.....
16693	Spinal cord	8.80	6.01	8.28	8.85	7.02	1.26	0.57	.....	.....
16695	Tendons.............................	31.93	10.23	1.52	14.10	11.26	2.22	0.62	0.39	3.71
17168	(Fat extracted with ether)	4.81	.....	.....	0.23	.....	.....	.....	6.65	.....
17177	Hoofs	63.44	7.14	1.35	14.63	.....	.....	.....	.....	1.46

PIG No. 2.—TAMWORTH.

16719	Bones	50.71	7.10	0.45	6.18	5.45	0.22	0.51	0.07	49.98
16719A	(Fat extracted with ether)	14.83	(0.68)	.....	0.34	.....	.....	.....	.....	(0.59)
16720	Marrow	2.39	6.91	0.44	10.56	8.98	0.56	1.02	.....	.....
17146	(Fat extracted with ether)	15.53	.....	.....	0.14	.....	.....	.....	0.31	.....
16717	Skin	34.28	7.08	4.04	14.88	7.75	4.21	2.92	0.34	1.89
17132	(Fat extracted with ether)	14.27	.....	.....	0.16	.....	.....	.....	0.91	.....
16722	Spinal cord..........................	8.88	6.41	1.65	9.63	7.30	1.12	1.21	.....	2.54
17166	(Fat extracted with ether)	32.24	.....	.....	0.44	.....	.....	.....	9.15	.....
16724	Tendons	33.67	9.01	0.83	14.86	11.49	2.05	1.32	0.31	3.24
17171	(Fat extracted with ether)	2.55	.....	.....	0.31	.....	.....	.....	.....	.....
17178	Hoofs	60.73	6.92	1.01	14.65	.....	.....	.....	.....	1.61

PIG No. 3.—CHESTER WHITE.

16630	Bones	45.56	6.46	0.45	6.94	6.04	0.17	0.73	0.13	47.61
16630A	(Fat extracted with ether)	16.97	(0.62)	.....	0.34	.....	.....	.....	.....	(0.37)
16631	Marrow	4.64	.....	.....	7.16	6.46	0.28	0.42	.....	.....
16633	Skin	28.05	3.37	3.98	15.02	2.56	3.46	9.00	0.35	1.88
16635	Spinal cord *a*	.....	.....	.....	.....	.....	.....	.....	.....	.....
16637	Tendons	34.22	13.03	0.84	14.10	11.34	1.53	1.23	0.81	2.48
17179	Hoofs	65.38	7.18	1.07	14.74	.....	.....	.....	.....	1.36

PIG No. 4.—POLAND CHINA.

16663	Bones	50.37	4.90	0.93	6.74	5.20	0.56	0.98	1.05	50.13
16663A	(Fat extracted with ether)	9.40	(0.62)	.....	0.21	.....	.....	.....	.....	(0.32)
16665	Marrow	5.28	6.66	0.43	8.28	7.12	0.42	0.14	.....	.....
16659	Skin	33.41	6.93	3.63	14.86	4.24	5.48	5.14	0.74	1.89
16661	Spinal cord..........................	12.25	6.58	10.15	9.12	7.16	1.40	0.50	.....	4.57
16662	(Fat extracted with ether)	23.32	.....	.....	0.17	.....	.....	.....	4.70	.....
16664	Tendons	35.22	9.48	0.67	14.63	11.24	2.50	0.89	0.29	4.21
17170	(Fat extracted with ether)	4.77	.....	.....	0.11	.....	.....	.....	.....	.....
17180	Hoofs	56.55	7.19	1.06	14.80	.....	.....	.....	.....	1.43

a Lost.

marrow, skin, spinal cord, tendons, and hoofs.

PIG No. 1.—BERKSHIRE.

Per cent original material.

Water.			Fat.			Nitrogen.		Nitrogenous substances.						
In preparing sample.	In original material.	Total.	In preparing sample.	In original material.	Total.	Lecithin.	Total.	Of proteids insoluble in hot water.	Precipitated by bromin.	Of flesh bases.	Proteids insoluble in hot water.	Gelatinoids.	Flesh bases.	Total.	Ash.	Total.
35.93	3.01	38.94	11.40	0.27	11.67	0.44	3.26	2.80	0.06	0.40	17.50	0.38	1.25	19.13	26.12	95.86
.....	(0.07)	.....	.....	.....	.....	.....	0.03	.....	.....	.....	.....	.....	.....	.....	(0.03)	.....
14.06	0.30	14.36	81.50	0.01	81.51	.....	0.37	0.32	0.03	0.02	2.00	0.19	0.06	2.25	.....	98.12
.....	.....	.....	.....	.....	.....	0.46	0.01	.....	.....	.....	.....	.....	.....	.....	.....	.....
47.17	3.07	50.24	15.90	1.21	17.11	0.12	5.55	4.04	1.07	0.44	25.25	6.00	1.37	33.31	0.63	{108.44 / 101.20}
.....	.....	.....	.....	.....	.....	0.29	0.024	.....	.....	.....	.....	.....	.....	.....	.....	.....
						0.41										
65.17	0.53	65.70	26.03	0.73	26.76	.....	0.78	0.62	0.11	0.05	3.88	0.69	0.16	4.73	.....	97.19
55.16	3.27	58.43	12.91	0.49	13.40	0.13	4.50	3.59	0.71	0.20	22.44	4.44	0.62	27.50	1.18	100.51
.....	.....	.....	.....	.....	.....	0.32	0.01	.....	.....	.....	.....	.....	.....	.....	.....	.....
						0.45										
36.56	4.53	41.09	.....	.....	0.86	.....	9.28	.....	.....	.....	.....	.....	.....	58.00	0.93	100.88

PIG No. 2.—TAMWORTH.

34.46	3.60	38.06	14.83	0.23	15.06	0.04	3.13	2.76	0.11	0.26	17.25	0.60	0.81	18.75	25.35	97.22
.....	(0.10)	.....	.....	.....	.....	.....	0.05	.....	.....	.....	0.31	.....	.....	0.31	(0.09)	0.31
											17.56			19.06		97.53
13.14	0.17	13.31	84.47	0.01	84.48	.....	0.25	0.22	0.01	0.02	1.38	0.06	0.06	1.50	.....	99.29
.....	.....	.....	.....	.....	.....	0.05	0.02	.....	.....	.....	.....	.....	.....	.....	.....	.....
52.95	2.43	55.38	12.77	1.38	14.15	0.12	5.10	2.66	1.44	1.00	16.62	9.00	3.12	28.74	0.65	{98.15 / 98.92}
.....	.....	.....	.....	.....	.....	0.13	0.02	.....	.....	.....	.....	.....	.....	.....	.....	.....
						0.25										
45.88	0.57	46.45	45.24	0.15	45.39	.....	0.86	0.65	0.10	0.11	4.06	0.63	0.34	5.03	0.23	97.10
						2.95	0.14									
58.52	3.03	61.55	7.81	0.28	8.09	0.10	5.00	3.87	0.69	0.44	24.19	4.31	1.37	29.87	1.09	100.60
.....	.....	.....	.....	.....	.....	.....	0.01	.....	.....	.....	.....	.....	.....	.....	.....	.....
30.27	4.20	43.47	.....	.....	0.61	.....	8.90	.....	.....	.....	.....	.....	.....	55.63	0.98	100.69

PIG No. 3.—CHESTER WHITE.

37.47	2.94	40.41	16.97	0.21	17.18	0.06	3.16	2.75	0.08	0.33	17.19	0.50	1.03	18.72	21.60	98.00
.....	(0.10)	.....	.....	.....	.....	.....	0.06	.....	.....	.....	0.38	.....	.....	0.38	(0.06)	0.38
											17.57			19.10		98.38
15.50	.....	.....	79.86	.....	.....	.....	0.33	0.30	0.01	0.02	1.88	0.06	0.06	2.00	.....	97.36
40.78	.....	40.78	31.17	.....	31.17	0.10	4.22	0.72	0.97	2.53	4.50	6.06	7.89	18.45	0.53	90.93
55.66	4.46	60.12	10.12	0.29	10.41	0.28	4.82	3.88	0.52	0.42	24.25	3.25	1.31	28.81	0.85	100.19
34.62	4.69	39.31	.....	.....	.....	0.70	.....	.....	.....	9.64	.....	.....	.....	60.25	0.89	101.15

PIG No. 4.—POLAND CHINA.

40.23	2.47	42.70	9.40	0.47	9.87	0.53	3.40	2.62	0.28	0.50	16.38	1.75	1.56	19.09	25.25	97.51
.....	(0.06)	.....	.....	.....	.....	.....	0.02	.....	.....	.....	0.13	.....	.....	0.13	(0.03)	0.13
											16.51			19.82		97.64
16.39	0.35	16.74	78.93	0.02	78.95	.....	0.44	0.41	0.02	0.01	2.56	0.13	0.03	2.72	.....	97.81
44.49	2.32	46.81	22.10	1.21	23.31	0.25	4.97	1.42	1.83	1.72	8.87	11.44	5.37	25.68	0.63	{88.62 / 96.43}
46.05	0.81	46.86	41.70	1.24	42.94	.....	1.12	0.88	0.17	0.07	5.50	1.06	0.22	6.78	0.56	97.14
.....	.....	.....	.....	.....	.....	1.10	0.04	.....	.....	.....	.....	.....	.....	.....	.....	.....
53.34	3.34	56.68	11.44	0.24	11.68	0.10	5.15	3.96	0	88	0.31	24.75	5 50	0.97	31.22	1.48	101.06
.....	.....	.....	.....	.....	.....	.....	0.005	.....	.....	.....	.....	.....	.....	.....	.....	.....
43.45	4.07	47.52	.....	.....	0.60	.....	8.37	.....	.....	.....	.....	.....	.....	52.31	0.81	101.24

TABLE NO. 5.—*Analytical data for bones*.

PIG No. 5.—DUROC JERSEY.

Serial No.	Names of parts.	Air-dry sample, per cent of original material.	Per cent air-dry material.							
			Water.	Fat.	Total	Nitrogen.			Lecithin.	Ash.
						Of proteids insoluble in hot water.	Precipitated by bromin.	Of flesh bases.		
16000	Bones	53.64	4.69	0.86	6.77	5.93	0.17	0.67	0.93	49.70
16600A	(Fat extracted with ether)	11.58	7.30		0.34					(0.61)
16601	Marrow	0.25	6.73	0.26	8.26	7.30	0.45	0.51		
16603	Skin	a 20.35	5.08	5.46	15.13	2.22	6.91	6.00	0.33	1.82
16605	Spinal cord	18.07	6.54	0.07	9.26	7.58	0.84	0.84		
16607	Tendons b									
17181	Hoofs	70.03	7.33	1.05	14.77					1.22

PIG No. 6.—DUROC JERSEY.

16748	Bones	51.74	5.79	0.31	6.40	5.39	0.31	0.76		50.36
16748A	(Fat extracted with ether)	17.48	(0.43)		0.21					(0.23)
16751	Marrow c		4.83	0.27	8.14	6.89	0.50	0.69		
16746	Skin	a 34.21	7.78	3.63	14.83	6.32	2.81	5.70	0.18	1.83
16749	Spinal cord c				9.97	7.72	1.12	1.13		
17162	(Fat extracted with ether)	34.91			0.20				8.12	
16753	Tendons	34.51	9.78	0.21	15.05	11.71	2.13	1.21	0.24	2.79
17173	(Fat extracted with ether)	4.11			0.12					
17182	Hoofs	58.83	5.97	1.18	14.55					1.73

PIG No. 7—DUROC JERSEY.

16775	Bones	53.50	6.52	0.06	6.44	5.59	0.17	0.68	0.15	50.59
16775A	(Fat extracted with ether)	12.88	3.63		0.32				2.72	
16776	Marrow c		6.35	0.48	7.02	6.32	0.56	0.14		
16778	Skin	39.18	9.80	3.78	14.60	7.84	4.13	2.72	0.20	2.00
17148	(Fat extracted with ether)	11.95			0.18				0.55	
16780	Spinal cord	12.76	6.43	0.76	9.55	7.58	1.20	0.71		
17167	(Fat extracted with ether)	50.36			0.37				5.50	
16782	Tendons	31.94	11.65	0.47	14.43	11.49	1.96	0.98	0.36	2.71
17183	Hoofs	60.06	6.96		14.83					

PIG No. 8.—YORKSHIRE.

16804	Bones	50.51	11.16	0.68	6.00	5.73	0.28	0.59	0.40	50.08
16804A	(Fat extracted with ether)	13.74	(0.53)		0.29					(0.26)
16805	Marrow	4.41	6.21	0.19	7.75	6.85	0.79	0.11		
17153	(Fat extracted with ether)	29.28			0.04				0.22	
16807	Skin c									
16809	Spinal cord c									
17101	(Fat extracted with ether)	30.30			0.27				5.29	
16811	Tendons		11.65	0.67	13.96	12.29	1.43	0.24	0.24	2.90
17184	Hoofs	51.03	6.44	0.96	14.65					1.39

a Fat-free and water-free.   b Sample lost.   c Lost.

marrow, skin, spinal cord, tendons, and hoofs—Continued.

PIG No. 5.—DUROC JERSEY.

Water.			Fat.			Nitrogen.						Nitrogenous substances.					
In preparing sample.	In original material.	Total.	In preparing sample.	In original material.	Total.	Lecithin.	Total.	Of proteids insoluble in hot water.	Precipitated by bromin.	Of flesh bases.	Proteids insoluble in hot water.	Gelatinoids.	Flesh bases.	Total.	Ash.	Total.
34.78	2.52	36.45	1.58	0.46	12.89	0.50	3.03	3.18	0.09	0.36	19.88	0.56	1.12	21.56	26.66	97.56
	0.85						0.04				0.25			0.25	(0.07)	0.25
											20.13			21.81		97.81
12.80	0.42	13.22	80.95	0.02	80.97		0.52	0.46	0.03	0.03	2.88	0.19	0.09	3.16		97.35
		35.49			38.16	0.09	3.99	0.59	1.82	1.58	3.69	11.38	4.93	20.00	0.48	94.13
58.32	1.18	59.50	23.61	0.01	23.62		1.07	1.37	0.15	0.15	8.56	0.94	0.47	9.97		93.09
29.97	5.13	35.10					0.74		10.34					64.63	0.85	101.32

PIG No. 6.—DUROC JERSEY.

30.78	3.00	33.78	17.48	0.16	17.64		3.34	2.79	0.16	0.39	17.44	1.00	1.22	19.66	26.06	97.14
	(0.08)						0.04				0.25			0.25	(0.04)	0.25
											17.60			19.91		97.39
		45.20			20.59	0.06	5.07	2.16	0.96	1.95	13.50	6.00	6.08	25.58	0.03	92.00
						2.51	0.07									
56.24	3.38	59.62	9.25	0.07	9.32	0.08	5.20	4.04	0.74	0.42	25.25	4.63	1.31	31.19	0.96	101.09
							0.005									
41.17	3.51	44.68			0.69		8.56							53.50	1.02	99.89

PIG No. 7.—DUROC JERSEY.

33.62	3.49		12.88	0.35		0.08	3.45	2.99	0.09	0.37	18.09	0.56	1.15	20.40	27.07	97.81
	0.47	36.64			13.70	0.35	0.04				0.25			0.25	(0.04)	0.25
						0.43					18.94			20.65		98.06
46.55		50.39	14.27		15.76	0.08	5.76	3.07	1.62	1.07	19.19	10.13	3.34	32.66	0.78	99.09 / 99.58
	3.84			1.48		0.07										
						0.15										
20.02	0.82	20.84	67.22	0.10	67.32		1.22	0.97	0.16	0.09	6.06	1.00	0.28	7.34		95.50
						0.70	0.19									
54.19	3.72	57.91	13.87	0.15	14.02	0.11	4.61	3.67	0.63	0.31	22.94	3.94	0.97	27.85	0.87	100.65
39.94	4.18	44.12					8.91							55.69		

PIG No. 8.—YORKSHIRE.

35.75	5.64	41.39	13.74	0.34	14.08	0.20	3.33	2.89	0.14	0.30	18.07	0.87	0.94	19.88	25.30	100.65
	(0.07)						0.04				0.25			0.25	(0.04)	0.25
											18.32			20.13		100.90
14.02	0.27	14.29	81.57	0.01	81.58		0.34	0.30	0.03	0.01	1.88	0.19	0.03	2.10		97.97
						0.06	0.01									
						1.60	0.08									
48.97	3.29	52.26			0.49		7.48							46.75	0.71	100.21

TABLE NO. 6—*Revised analytical data.*

PIG No. 1.—BERKSHIRE.

[Per cents original material.]

Serial No.	Names of cuts and parts.	Water.	Fat.	Nitrogenous substances.				Lecithin.a	Ash.	Total.
				Proteids, insoluble in hot water.	Gelatinoids.	Flesh bases.	Total.			
	Meat:									
16667	American backs.....	32.27	57.69	7.00	0.50	0.91	8.41	0.15	0.51	99.03
16669	American bellies	37.27	51.93	7.00	0.56	1.22	8.78	0.14	0.55	98.67
16671	Short-cut hams.......	b 60.29	22.19	14.00	0.69	1.15	15.84	0.65	0.96	99.93
16673	New York shoulders.	b 54.97	20.01	11.25	0.81	1.56	13.62	0.15	0.80	98.64
16675	Four feet.............	61.28	16.83	12.19	4.69	2.34	19.22	0.61	0.82	98.76
16677	Spareribs............	52.54	29.10	13.44	1.13	1.19	15.70	0.35	1.00	98.75
16679	Tenderloins	68.06	8.78	18.56	0.50	1.06	20.12	0.49	1.17	98.62
16680	Neck bones..........	55.70	27.92	12.44	0.75	1.06	14.25	0.68	0.81	99.30
16682	Backbones	52.83	27.22	14.38	0.87	1.44	16.69	0.26	1.24	98.24
16684	Trimmings..........	b 29.68	62.00	5.19	0.69	1.03	6.91	0.11	0.41	99.11
16686	Tail	24.02	68.23	5.75	0.56	0.50	6.81	0.17	0.39	99.62
16690	Bones................	38.94	11.67	17.50	0.38	1.25	19.13	0.44	26.12	96.30
16691	Marrow	14.36	81.51	2.00	0.19	0.06	2.25	c 0.46	......	98.58
16688	Skin.................	50.24	17.11	25.25	6.69	1.37	33.31	0.41	0.83	101.70
16693	Spinal cord..........	65.70	26.76	3.88	0.69	0.16	4.73	d 1.47	e 0.40	97.19
16695	Tendons.............	58.43	13.40	22.44	4.44	0.62	27.50	0.45	1.18	100.96
17177	Hoofs	41.09	0.86	......	......	......	58.00	......	0.03	100.88

PIG No. 2.—TAMWORTH.

	Meat:									
16696	American backs.....	b 29.13	61.76	5.75	0.09	0.91	7.35	0.13	0.43	98.80
16698	American bellies	33.09	56.52	7.00	0.63	0.91	8.54	0.15	0.47	99.37
16700	Short-cut hams......	b 57.93	24.45	13.00	0.63	1.25	14.88	0.22	0.84	98.32
16702	New York shoulders.	b 35.07 + 20 29.98		11.25	0.87	1.19	13.31	0.24	0.79	99.30
16704	Four feet.............	b 58.66	21.23	11.63	3.38	2.96	17.97	0.51	0.86	99.23
16706	Spareribs............	49.20	33.88	11.56	1.31	1.40	14.27	0.25	0.93	98.53
16708	Tenderloins	65.52	13.51	17.13	0.56	0.91	18.60	0.91	1.06	99.60
16709	Neck bones..........	55.52	26.03	13.38	1.31	1.40	16.09	0.83	1.02	99.49
16711	Backbones	51.06	30.67	13.69	0.87	1.22	15.78	0.25	1.10	98.86
16713	Trimmings..........	28.85	62.67	5.31	0.81	1.03	7.15	0.10	0.43	99.20
16715	Tail	25.77	67.08	4.25	0.69	0.78	5.72	0.10	0.30	98.97
16719	Bones................	38.06	15.06	17.56	0.69	0.81	19.06	0.04	25.35	97.57
16720	Marrow	13.31	84.48	1.38	0.06	0.06	1.50	f 0.05	......	99.34
16717	Skin.................	55.38	14.15	16.62	9.00	3.12	28.74	0.25	0.65	99.17
16722	Spinal cord..........	46.45	45.39	4.06	0.63	0.34	5.03	f 2.95	0.23	100.05
16724	Tendons.............	61.55	8.09	24.19	4.31	1.37	29.87	0.10	1.09	100.70
17178	Hoofs	43.47	0.61	......	......	......	55.63	......	0.98	100.69

PIG No. 3.—CHESTER WHITE.

	Meat:									
16609	American backs	b 23.72	70.16	4.50	0.44	0.75	5.69	0.12	0.35	100.04
16611	American bellies	30.54	60.73	5.44	0.63	1.03	7.10	0.08	0.42	98.87
16613	Short-cut hams	53.15	30.99	11.13	0.63	1.97	13.73	0.35	0.80	99.02
16615	New York shoulders	b 49.16	37.62	9.44	0.87	1.34	11.65	0.28	0.71	99.42
16617	Four feet.............	b 53.05	26.74	9.88	8.38	1.06	17.32	0.19	0.84	98.04
16619	Spareribs	53.23	27.93	13.63	0.87	1.65	16.15	0.28	0.92	98.51
16621	Tenderloins	65.97	13.47	16.69	0.44	1.03	18.16	0.40	1.06	99.06
16622	Neck bones	53.50	29.52	12.19	0.75	1.31	14.25	0.25	0.87	98.39
16624	Backbones	50.30	31.65	13.38	0.56	1.34	15.28	0.23	1.05	98.00
16626	Trimmings..........	22.49	71.55?	3.25	0.50	0.84	4.59	0.07	0.31	90.01
16628	Tail	b 15.66	79.69	3.00	0.44	0.62	4.06	0.08	0.24	99.73
16630	Bones................	40.41	17.18	17.57	0.50	1.03	19.10	0.06	21.69	98.44
16631	Marrow	15.50	79.86	1.88	0.06	0.06	2.00	c 0.19	......	97.42
16633	Skin	b 40.78	31.17	4.50	6.06	7.89	18.45	0.10	0.53	91.03
16635	Spinal cord	e 48.27	e 41.21	e 5.61	e 0.86	e 0.30	e 6.77	c 1.47	e 0.40	498.12
16637	Tendons	60.12	10.41	24.25	3.25	1.31	28.81	0.28	0.85	100.47
17179	Hoofs	39.31	0.70	......	......	......	60.25	......	0.89	101.15

a Lecithin in extracted sample only, unless otherwise noted.
b Result of direct determination on original material. Other numbers in this column represent the sum of the per cent of water removed in the preparation of sample and the per cent of water remaining in the air-dry sample.
c In fat extract.
d In fat extract, calculated from averages for like cuts.
e Calculated from averages of like cuts.
f In residue and fat extract.

49

TABLE No. 6—*Revised analytical data*—Continued.

PIG No. 4.—POLAND CHINA.

[Per cents original material.]

Serial No.	Names of cuts and parts.	Water.	Fat.	Proteids, insoluble in hot water.	Gelatinoids.	Flesh bases.	Total.	Lecithin. a	Ash.	Total.
	Meat:									
16638	American backs	26.13	66.33	5.13	0.56	0.81	6.50	a 0.21	0.38	99.34
16640	American bellies	30.78	60.69	4.81	0.63	1.19	6.63	0.12	0.43	98.65
16642	Short-cut hams	54.78	30.12	10.69	0.81	1.28	12.78	0.23	0.76	96.67
16644	New York shoulders	b 51.72	33.74	7.81	0.63	0.94	9.38	0.30	0.59	95.73
16646	Four feet	50.66	31.32	11.00	4.19	0.97	16.16	0.15	0.91	99.20
16648	Spareribs	b 52.95	29.55	10.06	0.63	3.09	13.78	0.31	0.95	97.54
16650	Tenderloins	67.43	10.95	17.63	0.69	0.78	19.10	0.39	1.13	99.00
16651	Neck bones	55.23	28.27	10.25	1.13	1.75	13.13	0.45	0.01	97.99
16653	Backbones	b 51.26	30.98	12.94	0.81	1.34	15.09	0.27	1.05	98.65
16655	Trimmings	22.43	71.52	3.69	0.44	0.69	4.82	0.06	0.32	99.15
16657	Tail	16.50	77.77	3.44	0.56	0.66	4.66	0.00	0.31	99.33
16663	Bones	42.70	9.87	16.51	1.75	1.56	19.82	0.53	25.25	98.17
16665	Marrow	16.74	78.35	2.56	0.13	0.03	2.72	c 0.19		97.87
16659	Skin	46.81	23.31	8.87	11.44	5.37	25.68	0.25	0.63	96.68
16661	Spinal cord	48.86	42.94	5.50	1.06	0.22	6.78	1.10	0.56	100.24
16664	Tendons	56.68	11.68	24.75	5.50	0.97	31.22	0.10	1.48	101.10
17180	Hoofs	47.52	0.60				52.31		0.81	101.24

PIG No. 5.—DUROC JERSEY.

	Meat:									
16579	American backs	b 20.75	73.25	3.94	0.50	0.84	5.28	0.08	0.29	99.65
16581	American bellies	20.13	62.83	5.06	0.56	1.00	6.62	0.10	0.42	99.10
16583	Short-cut hams	50.45	35.94	8.38	1.00	0.50	9.88	0.03	0.71	97.01
16585	New York shoulders	b 44.16	43.74	7.63	0.63	1.09	9.35	0.10	0.55	97.90
16587	Four feet	b 54.16	26.19	10.19	2.19	2.90	15.28	a 0.32	0.76	96.39
16589	Spareribs	54.09	26.90	14.06	0.81	1.53	16.40	0.12	1.04	98.55
16591	Tenderloins	66.55	11.41	18.31	0.38	1.09	19.78	0.35	1.14	99.23
16592	Neck bones	52.48	30.31	12.50	0.75	1.40	14.65	0.20	0.94	98.58
16594	Backbones	48.77	29.40	16.38	0.81	1.65	18.84	0.41	1.23	98.65
16596	Trimmings	20.08	74.13	3.25	0.44	0.87	4.56	0.08	0.30	99.15
16598	Tail	11.54	84.62	2.00	0.31	0.62	2.93	0.07	0.20	99.36
16600	Bones	36.45	12.89	20.13	0.56	1.12	21.81	0.50	26.66	98.31
16601	Marrow	13.22	80.97	2.88	0.19	0.09	3.16	d 0.19		97.41
16603	Skin	b 35.49	38.16	3.69	11.38	4.93	20.00	0.09	0.48	94.22
16605	Spinal cord	59.50	23.62	8.56	0.94	0.47	9.97	1.47	a 0.40	93.09
16607	Tendons	a 59.05	a 11.15	a 23.97	a 4.35	a 1.09	a 29.41	a 0.19	a 1.07	100.68
17181	Hoofs	35.10	0.74				64.63		0.85	101.32

PIG No. 6.—DUROC JERSEY.

	Meat:									
16725	American backs	20.32	73.63	3.75	0.44	0.72	4.91	0.20	0.29	99.35
16727	American bellies	34.52	58.97	4.38	0.56	0.59	5.53	0.12	0.22	99.36
16729	Short-cut hams	37.26	39.10	17.38	1.25	1.69	20.32	0.35	1.32	98.35
16731	New York shoulders	39.46	49.18	7.50	0.94	1.06	9.50	0.12	0.84	99.10
16733	Four feet	51.84	29.47	10.56	2.88	2.31	15.75	0.35	0.75	98.16
16735	Spareribs	49.84	31.95	13.63	1.13	1.09	15.85	0.33	1.10	99.07
16737	Tenderloins	63.86	14.81	17.31	0.63	0.87	18.81	0.56	1.05	99.09
16738	Neck bones	49.30	34.92	11.60	0.94	1.06	13.69	0.28	0.87	99.06
16740	Backbones	b 47.54	35.96	12.81	1.06	1.15	15.02	0.36	0.92	99.80
16742	Trimmings	16.51	78.78	3.06	0.38	0.50	3.94	0.07	0.25	99.54
16744	Tail	13.94	81.23	3.00	0.44	0.41	3.85	0.06	0.23	99.31
16748	Bones	33.78	17.84	17.69	1.00	1.22	19.91	a 0.31	26.06	97.39
16751	Marrow	a 14.57	a 81.13	a 2.29	a 0.14	a 0.66	a 2.29	d 0.19		97.99
16746	Skin	b 45.20	20.50	13.50	6.00	6.08	25.58	0.06	0.63	92.06
16749	Spinal cord	a 48.27	a 41.21	a 5.61	a 0.86	a 0.30	a 6.77	e 2.51	a 0.40	99.16
16753	Tendons	59.62	9.32	25.25	4.63	1.31	31.19	0.08	0.96	101.10
17182	Hoofs	44.68	0.69				53.50		1.02	99.89

a Calculated from averages of like cuts.
b Result of direct determination on original material. Other numbers in this column represent the sum of the per cent of water removed in the preparation of sample and the per cent of water remaining in the air-dry sample.
c In fat extract.
d In residue and fat extract, calculated from averages of like cuts.
e In residue and fat extract.

TABLE No. 6.—*Revised analytical data*—Continued.

PIG No. 7.—DUROC JERSEY.

[Per cents original material.]

Serial No.	Names of cuts and parts.	Water.	Fat.	Nitrogenous substances.				Lecithin.a	Ash.	Total.
				Proteids, insoluble in hot water.	Gelatinoids.	Flesh bases.	Total.			
	Meat:									
16754	American backs	20.23	73.95	4.06	0.56	0.50	5.12	0.48	0.29	100.07
16756	American bellies	a21.53	73.56	2.81	0.38	0.31	3.50	0.48	0.21	99.28
16758	Short-cut hams	a44.26	43.38	9.56	0.87	0.81	11.24	0.45	0.65	99.98
16760	New York shoulders.	44.62	42.65	9.06	0.94	0.87	10.87	0.62	0.91	99.67
16762	Four feet	55.98	25.25	12.38	2.63	1.53	16.54	0.36	0.78	98.91
16764	Spareribs	53.20	27.51	14.50	1.13	1.09	16.78	0.83	1.01	99.33
16766	Tenderloins	72.83	9.35	14.69	0.50	0.56	15.75	0.32	0.86	99.11
16767	Neck bones	52.10	30.92	12.38	1.13	0.62	14.13	0.55	0.94	98.64
16769	Backbones	53.09	28.07	13.88	1.06	1.19	16.13	0.44	1.02	98.75
16771	Trimmings	20.49	73.56	3.38	0.63	0.75	4.76	b0.11	0.28	99.20
16773	Tail	13.73	81.74	2.81	0.50	0.44	3.75	0.48	0.27	99.07
16775	Bones	36.64	13.70	18.94	0.56	1.15	20.65	0.43	27.07	98.49
16776	Marrow	c14.57	c81.13	c2.09	c0.14	c0.06	c2.29	d0.19		97.09
16778	Skin	50.30	15.75	19.19	10.13	3.34	32.66	0.15	0.78	99.73
16780	Spinal cord	20.84	67.32	6.06	1.00	0.28	7.34	b0.70	c0.40	96.20
16782	Tendons	57.91	14.02	22.94	3.94	0.97	27.85	0.11	0.87	100.76
17183	Hoofs	44.12	e0.67				55.69		c0.89	101.37

PIG No. 8.—YORKSHIRE.

	Meat:									
16783	American backs	28.55	62.53	5.94	1.06	0.62	7.62	0.32	0.41	99.43
16785	American bellies	33.70	56.22	6.75	1.25	0.66	8.66	0.25	0.48	99.40
16787	Short-cut hams	59.14	25.23	11.50	0.87	1.09	13.46	0.42	0.74	98.99
16789	New York shoulders.	a49.57	36.09	8.50	0.87	1.09	10.46	0.17	0.63	96.92
16791	Four feet	a57.47	30.80	5.13	1.63	1.53	8.29	0.08	0.41	97.11
16793	Spareribs	52.31	29.28	13.56	1.13	1.34	16.03	0.33	1.05	99.00
16795	Tenderloins	66.05	13.07	16.00	0.81	1.03	17.84	0.68	1.04	99.28
16796	Neck bones	55.13	26.71	13.31	1.00	1.15	15.46	0.54	0.99	98.83
16798	Backbones	50.65	29.61	14.13	1.25	1.28	16.66	0.78	1.18	98.88
16800	Trimmings	26.12	65.81	4.50	1.00	0.94	6.50	0.44	0.39	99.26
16802	Tail	18.50	74.37	4.88	1.06	0.41	6.35	0.55	0.33	100.10
16804	Bones	41.39	14.08	18.32	0.87	0.94	20.13	0.20	25.30	101.10
16805	Marrow	14.20	81.58	1.88	0.19	0.03	2.10	b0.06		98.03
16807	Skin	c46.33	c22.88	c13.09	c8.67	c4.59	c26.35	c0.19	c0.62	96.18
16809	Spinal cord	c48.27	c41.21	c5.61	c0.86	c0.30	c6.77	e1.47	c0.40	96.73
16811	Tendons	c59.05	c11.15	c23.97	c4.35	c1.09	c29.41	c0.19	c1.07	100.68
17184	Hoofs	52.20	0.49				46.75		0.71	100.21

a Result of direct determination on original material. Other numbers in this column represent the sum of the per cent of water removed in the preparation of sample and the per cent of water remaining in the air-dry sample.
b In residue and fat extract.
c Calculated from averages of like cuts.
d In fat extract, calculated from averages for like cuts.
e In fat extract.

51

TABLE No. 7.—*Data for the entire dressed animal; the head, leaf lard, and kidneys having been removed.*

PIG No. 1.—BERKSHIRE.

Serial No.	Names of parts.	Weight of parts—			Weight of each constituent.						
		From each cut.	Total.	Of entire pig.	Water.	Fat.	Nitrogenous substances.			Lecithin.	Ash.
							Proteids, insoluble in hot water.	Gelatinoids.	Flesh bases.	Total.		
		Grams.	*Grams.*	*Per cent.*	*Grams.*	*Grams.*	*Grams.*	*Grams.*	*Grams.*	*Grams.*	*Grams.*	*Grams.*
	Meat (fat and lean):											
16667	Backs	14,767.9			4,765.6	8,519.6	1,033.8	73.8	134.4	1,242.0	22.15	75.3
16669	Bellies	8,230.6			3,067.5	4,274.2	576.1	46.1	100.4	722.6	11.52	45.3
16671	Hams	9,407.9			5,672.0	2,067.6	1,317.2	64.9	108.2	1,490.3	61.15	90.3
16673	Shoulders	8,448.2			4,644.2	2,450.6	950.4	68.4	131.8	1,150.6	12.67	75.3
16675	Feet	325.3			199.3	54.7	39.6	15.3	7.6	62.5	1.98	2.7
16677	Spareribs	1,683.8			884.7	490.0	226.4	19.0	20.0	265.4	5.89	16.8
16679	Tenderloins	470.8			320.4	41.3	87.4	2.3	5.0	94.7	2.31	5.5
16680	Neck bones	493.2			274.7	137.7	61.4	3.7	5.2	70.3	3.35	4.0
16682	Back bones	704.0			371.9	191.7	101.3	6.1	10.1	117.5	1.83	8.7
16684	Trimmings	7,021.5			2,084.0	4,353.1	364.4	48.4	72.3	485.1	7.72	28.8
16686	Tail	291.7			70.1	199.1	16.8	1.6	1.5	19.9	0.50	1.1
	Total for meats		51,844.9	88.19	22,354.4	22,799.6	4,774.8	349.6	596.5	5,720.9	131.07	353.7
16690	Bones (less marrow)		4,374.7	7.44	1,703.6	510.6	765.6	16.6	54.7	836.9	19.25	1,142.6
16691	Marrow		69.7	0.12	10.0	56.8	1.4	0.2	0.1	1.7	a0.32	
16668	Skin		2,232.5	3.80	1,121.6	391.9	563.7	149.4	30.6	743.7	9.15	14.1
16693	Spinal cord		55.7	0.09	36.6	14.9	2.2	0.4	0.1	2.7	b0.82	c0.2
16695	Tendons		159.5	0.27	93.2	21.4	35.8	7.1	1.0	43.9	0.72	1.9
17177	Hoofs		52.6	0.09	21.6	0.4				30.5		0.5
	Total weights		58,789.6		25,341.0	23,785.6	6,143.5	523.3	683.0	7,654.9	101.33	151.3
	Total per cents of original material				43.10	40.46	10.45	0.89	1.16	13.02	0.27	2.57

a In fat extract. *b* In residue and fat extract, calculated from averages of like cuts. *c* Calculated from averages of like cuts.

52

TABLE No. 7.—*Data for the entire dressed animal; the head, leaf lard, and kidneys having been removed*—Continued.

PIG No. 2.—TAMWORTH.

Serial No.	Name of parts.	Weight of parts—		Of entire pig.	Weight of each constituent.						
		From each cut.	Total.		Water.	Fat.	Nitrogenous substances.			Total.	Lecithin.	Ash.
							Proteids, insoluble in hot water.	Gelatinoids.	Flesh bases.			
		Grams.	*Grams.*	*Per cent.*	*Grams.*	*Grams.*	*Grams.*	*Grams.*	*Grams.*	*Grams.*	*Grams.*	*Grams.*
	Meat (fat and lean):											
16696	Backs	17,201.7			5,010.9	10,624.5	989.2	118.7	156.6	1,264.5	22.36	74.0
16698	Bellies	8,281.8			2,790.0	4,680.9	579.7	52.2	75.4	707.3	12.42	38.9
16700	Hams	10,110.2			5,856.8	2,471.9	1,314.2	63.7	120.4	1,504.3	22.24	84.9
16702	Shoulders	8,256.6			4,546.9	2,475.0	928.8	71.8	98.2	1,098.8	19.82	65.2
16704	Feet	452.3			265.3	96.0	52.6	15.3	13.4	81.3	2.31	3.9
16706	Spareribs	1,575.7			775.3	533.9	182.2	20.6	22.1	224.9	3.94	14.7
16708	Tenderloins	528.2			346.1	71.4	90.5	3.0	4.8	98.3	4.81	5.6
16709	Neck bones	459.5			253.1	110.6	61.5	6.0	6.4	73.9	3.81	4.7
16711	Backbones	860.2			430.2	263.8	117.8	7.5	10.5	135.8	2.05	9.5
16713	Trimmings	6,979.4			2,013.6	4,374.0	370.6	56.5	71.9	499.0	0.70	30.0
16715	Tail	590.5			154.5	402.2	25.5	4.1	4.7	34.3	0.60	1.8
	Total for meats		55,305.1	86.50	22,453.7	26,113.2	4,712.6	419.4	590.4	5,722.4	95.06	331.2
16719	Bones (less marrow)		5,231.8	8.18	1,991.2	787.9	918.7	36.1	42.4	997.2	2.09	1,326.1
16720	Marrow		133.9	0.21	17.8	113.1	0.1	0.1	0.1	2.0	0.07	
16717	Skin		3,011.6	4.71	1,067.9	426.2	500.3	271.1	94.0	805.4	7.53	19.6
16722	Spinal cord		60.8	0.09	28.2	27.6	2.5	0.4	0.2	3.1	1.79	0.1
16724	Tendons		138.7	0.21	85.4	11.2	33.6	0.0	1.9	41.5	0.14	1.5
17178	Hoofs		64.5	0.10	28.0	0.4				35.0		0.6
	Total weights		63,946.4		26,272.2	27,479.6	6,160.6	733.1	729.0	7,667.5	106.68	1,681.1
	Total per cents of original material			100.00	41.09	42.97	9.65	1.15	1.14	11.99	0.17	2.63

PIG No. 3.—CHESTER WHITE.

	Meat (fat and lean):											
16609	Backs	15,217.0			3,909.9	10,076.9	684.9	66.9	114.1	806.0	18.26	53.3
16611	Bellies	8,966.0			2,738.4	5,445.1	487.8	56.5	92.4	636.7	7.17	37.7
16613	Hams	7,765.3			2,406.7	2,406.7	864.4	48.9	153.6	1,066.3	27.19	62.1
16615	Shoulders	8,408.4			4,163.1	3,185.9	799.4	73.7	113.5	986.6	23.71	60.1
16617	Feet	209.1			142.8	72.0	20.6	17.2	2.8	40.6	0.49	2.3

53

16619	Spareribs	1,012.8		539.1	282.9	138.1	8.8	16.7	163.6	2.84	9.3
16021	Tenderloins	453.6		290.5	61.1	75.7	2.0	4.7	82.4	1.81	4.8
16022	Neck bones	427.1		228.5	126.1	52.1	3.2	5.6	60.9	1.07	3.7
16024	Backbones	584.5		249.5	185.0	78.2	3.3	7.8	89.3	1.34	6.2
16026	Trimmings	6,398.9		1,439.1	4,578.2	208.0	32.0	53.7	293.7	4.48	19.8
16628	Tail	633.9		99.3	505.2	19.0	2.8	3.9	25.7	0.51	1.5
	Total for meats	50,198.1	87.94	17,636.5	27,525.1	3,434.2	315.3	568.2	4,317.8	88.87	260.8
16630	Bones (less marrow)	3,545.2	6.21	1,432.6	609.1	623.0	17.7	30.5	677.2	2.13	769.0
16631	Marrow	43.6	0.08	6.8	34.8	0.03		0.03	0.9	a 0.08	
16033	Skin	3,150.2	5.52	1,284.7	983.0	141.8	190.9	248.6	581.3	3.15	16.7
16635	Spinal cord	39.0	0.07	(18.8)	(16.1)	(2.2)	(0.3)	(0.1)	(2.6)	b 0.57	0.2
16637	Tendons	67.2	0.12	40.4	7.0	16.3	2.2	0.9	10.4	0.19	0.6
17179	Hoofs	37.0	0.06	14.5	0.3				22.3		0.3
	Total weights	57,080.3		20,434.3	29,174.4	4,218.3	526.43	854.33	5,621.5	94.99	1,047.6
	Total per cent of original material		100.00	35.80	51.11	7.39	0.92	1.50	9.85	0.17	1.84

PIG No. 4.—POLAND CHINA.

	Meat (fat and lean):										
16638	Backs	16,830.2		4,397.9	11,162.4	863.4	94.3	136.3	1,094.0	(35.34)	64.0
16640	Bellies	10,151.8		3,124.7	6,161.0	488.3	64.0	120.8	673.1	12.18	43.7
16642	Hams	10,636.0		5,696.4	3,203.6	1137.1	88.2	136.1	1,359.4	4.48	80.8
16644	Shoulders	9,735.5		5,035.2	3,284.6	760.4	61.3	91.5	913.2	29.21	57.4
16646	Feet	466.6		236.4	146.2	51.3	19.6	4.5	75.4	0.70	4.5
16648	Spareribs	1,522.4		806.2	449.9	153.2	9.6	47.0	209.8	4.72	14.5
16650	Tenderloins	419.8		283.2	46.0	74.0	2.9	3.3	80.2	1.64	4.7
16651	Neck bones	537.9		297.1	152.1	55.1	6.1	9.4	70.6	2.42	5.0
16653	Backbones	636.6		326.6	197.2	82.4	5.2	8.5	99.1	1.72	6.7
16655	Trimmings	8,574.3		1,923.2	6,132.2	316.4	37.7	59.2	413.3	5.13	27.4
16657	Tail	671.6		110.8	522.3	23.1	3.8	4.4	31.3	0.60	2.1
	Total for meats	60,182.7	90.67	22,367.7	31,457.5	4,004.7	390.7	621.0	5,016.4	98.14	310.5
16663	Bones (less marrow)	3,519.0	5.30	1,502.6	347.7	581.1	61.6	54.9	697.6	18.65	888.7
16605	Marrow	72.0	0.11	12.1	56.4	0.1		0.02	1.92	a 0.14	
16659	Skin	2,406.8	3.63	1,126.6	561.0	213.5	275.4	129.3	818.2	6.02	15.3
16661	Spinal cord	51.1	0.08	25.0	21.9	2.8	0.5	0.1	3.4	0.56	0.3
16664	Tendons	96.0	0.14	54.4	11.2	23.8	5.3	0.9	30.0	0.10	1.4
17180	Hoofs	44.9	0.07	21.8	0.3				23.5		0.4
	Total weights	66,372.5		25,109.7	32,456.0	4,827.7	733.6	806.22	6,408.3	123.61	1,216.5
	Total per cent of original material		100.00	37.83	48.90	7.27	1.11	1.21	9.66	0.19	1.83

a In fat extract, calculated from averages for like cuts.   b In residue and fat extract, calculated from averages of like cuts.

54

TABLE No. 7.—*Data for the entire dressed animal; the head, leaf lard, and kidneys having been removed*—Continued.

PIG No. 5.—DUROC JERSEY.

Serial No.	Names of parts.	Weight of parts—		Of entire pig.	Weight of each constituent.						
		From each cut.	Total.		Water.	Fat.	Proteids, insoluble in hot water.	Nitrogenous substances.			Ash.
								Gelatinoids.	Flesh bases.	Total.	Lecithin.	
		Grams.	*Grams.*	*Per cent.*	*Grams.*	*Grams.*	*Grams.*	*Grams.*	*Grams.*	*Grams.*	*Grams.*	*Grams.*
	Meat (fat and lean):											
16579	Backs	10,709.2			3,457.2	12,238.5	658.4	83.5	140.4	882.3	13.37	48.5
16581	Bellies	10,189.1			2,988.1	6,402.1	515.6	57.1	101.9	674.6	10.10	42.8
16583	Hams	8,396.1			4,235.8	3,017.9	703.6	84.0	42.0	829.6	2.52	59.6
16585	Shoulders	7,971.0			3,520.0	3,486.6	608.2	50.8	80.9	745.0	7.97	43.8
16587	Feet	200.1			108.4	52.4	20.4	4.4	5.8	30.6	(0.64)	1.5
16589	Spareribs	1,185.4			646.6	321.6	168.1	9.7	18.3	196.1	1.43	12.4
16591	Tenderloins	348.5			231.9	39.8	63.8	1.3	3.8	68.9	1.22	4.0
16592	Neck bones	488.5			256.4	148.1	61.1	3.7	6.8	71.6	0.98	4.6
16594	Backbones	702.3			342.5	200.5	115.1	5.7	11.6	132.4	2.88	8.8
16596	Trimmings	8,155.7			1,637.7	6,046.0	265.1	35.9	70.9	371.9	6.52	24.5
16608	Tail	590.6			68.9	504.9	11.9	1.9	3.7	17.5	0.42	1.2
	Total for meats		54,952.5	88.03	17,483.5	32,464.4	3,191.3	338.0	492.1	4,021.4	48.14	251.5
16600	Bones (less marrow)		3,697.3	5.92	1,317.7	476.6	744.3	20.7	41.4	806.4	18.49	985.7
16601	Marrow		67.2	0.11	8.9	54.4	1.9	0.1	0.1	2.1	*a*0.12	
16603	Skin		3,590.2	5.75	1,274.2	1,370.1	132.5	408.0	177.0	718.1	3.23	17.2
16605	Spinal cord		25.7	0.04	15.3	6.1	2.2	0.24	0.12	2.56	*a*0.38	*b*0.1
16607	Tendons		59.6	0.10	35.2	6.7	*b*14.3	*b*2.6	*b*0.6	*b*17.5	*b*0.11	*b*0.6
17181	Hoofs		31.6	0.05	11.1	0.2				20.4		0.3
	Total weights		62,424.1		20,165.9	34,378.5	4,086.5	772.4	712.4	5,611.5	70.47	1,265.4
	Total per cents of original material			100.00	32.32	55.07	6.55	1.24	1.14	8.90	0.11	2.01

PIG No. 6.—DUROC JERSEY.

	Meat (fat and lean):											
16725	Backs	19,104.9			3,900.4	14,133.2	720.0	84.5	138.2	942.7	38.39	55.7
16727	Bellies	14,101.6			4,867.9	8,315.9	617.7	79.0	83.2	779.9	16.92	31.0
16729	Hams	11,120.6			4,144.5	4,348.8	1,933.0	139.6	188.0	2,260.0	38.92	146.8
16731	Shoulders	9,171.7			3,619.2	4,511.0	687.9	86.2	97.2	871.3	11.01	77.0
16733	Feet	493.7			256.0	145.5	52.2	14.2	11.4	77.8	1.73	3.7

16735	Spareribs	1,214.3		605.2	388.0	165.5	13.7	13.2	192.4	4.01	13.4
16737	Tenderloins	421.3		269.1	62.4	72.9	2.7	3.7	70.3	2.36	4.4
16738	Neck bones	635.6		313.4	222.0	74.3	6.0	6.7	87.0	1.78	5.5
16740	Backbones	809.7		384.9	201.2	103.7	8.6	9.3	121.6	2.91	7.5
16742	Trimmings	10,679.6		1,763.1	8,413.4	326.8	40.6	53.4	420.8	7.48	26.7
16744	Tail	1,070.7		140.3	869.7	32.1	4.7	4.4	41.2	0.64	2.5
	Total for meats	68,913.7		20,273.0	41,701.1	4,786.1	479.2	608.7	5,874.0	126.15	374.2
16748	Bones (less marrow)	3,564.7		1,704.2	628.8	630.6	35.7	43.5	709.8	(12.05)	928.9
16751	Marrow	79.0		b11.5	b64.1	b1.6	b0.2	b0.05	b1.75	a0.15	
16746	Skin	3,035.3		1,372.0	625.7	409.8	182.1	184.6	176.5	1.82	19.1
16749	Spinal cord	55.0		b26.6	b22.7	b3.1	b0.5	b0.2	b3.8	c1.38	b0.2
16753	Tendons	92.5		55.2	8.6	23.4	4.3	1.2	28.9	0.07	0.9
17182	Hoofs	59.2		26.4	0.4				31.7		0.6
	Total weights	75,799.4		22,968.9	43,050.7	5,854.6	701.9	838.25	7,426.45	141.62	1,323.9
	Total per cents of original material		100.00	30.31	56.81	7.73	0.93	1.11	9.80	0.19	1.75

PIG No. 7.—DUROC JERSEY.

	Meat (fat and loan):										
16754	Backs	16,807.7		3,400.3	12,429.3	682.4	94.1	84.0	860.5	80.68	48.8
16756	Bellies	12,404.8		2,670.9	9,125.0	348.6	47.1	38.5	434.2	50.54	26.1
16758	Hams	9,746.2		4,313.7	4,227.0	931.8	84.8	78.9	1,095.5	43.86	63.4
16760	Shoulders	8,043.4		3,589.0	3,430.2	728.7	75.8	70.0	874.3	49.87	73.2
16762	Feet	290.8		62.8	73.4	36.0	7.7	4.5	46.2	1.05	2.3
16764	Spareribs	1,114.1		592.7	306.5	162.2	12.6	12.2	187.0	9.25	11.3
16766	Tenderloins	333.3		242.8	31.2	40.0	1.7	1.8	52.6	1.07	2.9
16767	Neck bones	611.4		318.5	189.1	75.1	6.9	3.8	80.4	3.36	2.5
16769	Backbones	785.8		417.2	220.6	109.1	8.3	9.4	126.8	3.46	8.0
16771	Trimmings	10,029.2		2,055.0	7,377.1	339.0	63.2	75.2	477.4	11.03	28.1
16773	Tail	672.8		92.4	549.2	18.9	3.4	3.0	25.3	3.23	1.8
	Total for meats	60,839.5		17,755.3	37,960.2	3,481.4	405.4	381.4	4,268.2	266.40	271.7
16775	Bones (less marrow)	3,426.6		1,255.3	409.4	648.9	19.2	30.4	707.5	14.73	927.4
16776	Marrow	76.1		b11.1	b61.7	b1.6	b0.1	b0.05	b1.75	a0.15	
16778	Skin	3,140.8		1,587.2	496.1	604.4	319.1	105.2	1,028.7	4.72	24.6
16780	Spinalcord	56.2		11.7	37.8	3.4	0.6	0.2	4.2	0.39	b0.2
16782	Tendons	78.0		45.2	10.9	17.9	3.1	0.8	21.8	0.00	0.7
17183	Hoofs	52.0		22.9	0.3				29.0		0.5
	Total weights	67,677.6		20,688.7	39,036.4	4,757.6	747.5	527.05	6,061.15	286.48	1,225.1
	Total per cents of original material		100.00	30.58	57.68	7.03	1.10	0.78	8.96	0.42	1.81

a In fat extract, calculated from averages for like cuts.   b Calculated from averages of like cuts.   c In residue and fat extract.

56

TABLE No. 7.—*Data for the entire dressed animal; the head, leaf lard, and kidneys having been removed*—Continued.

PIG No. 8.—YORKSHIRE.

Serial No.	Names of parts.	Weight of parts—			Weight of each constituent.							
		From each cut.	Total.	Of entire pig.	Water.	Fat.	Nitrogenous substances.				Lecithin.	Ash.
							Proteids, insoluble in hot water.	Gelati- noids.	Flesh bases.	Total.		
		Grams.	*Grams.*	*Per cent.*	*Grams.*	*Grams.*	*Grams.*	*Grams.*	*Grams.*	*Grams.*	*Grams.*	*Grams.*
	Meat (fat and lean):											
16783	Backs	18,458.1			5,269.6	11,541.9	1,096.3	195.7	114.4	1,406.4	57.22	75.7
16785	Bellies	9,502.2			3,231.0	5,376.0	645.4	119.5	63.1	828.0	23.91	45.9
16787	Hams	10,682.4			6,317.6	2,695.0	1,228.4	92.9	116.5	1,437.8	44.87	79.0
16789	Shoulders	11,124.8			5,514.6	4,015.1	945.6	96.8	121.3	1,163.7	18.91	70.1
16791	Feet	673.3			387.0	207.8	34.5	11.0	10.3	55.8	0.54	2.8
16793	Spareribs	1,774.0			928.0	519.4	240.6	20.1	23.8	284.5	5.85	18.6
16795	Tenderloins	632.5			421.0	82.7	101.2	5.1	6.5	112.8	4.30	6.6
16796	Neck bones	770.9			425.0	205.9	102.6	7.7	8.9	119.2	4.16	7.6
16798	Backbones	1,100.0			557.2	325.8	155.4	13.8	14.1	183.3	8.58	13.0
16800	Trimmings	7,743.6			2,022.0	5,096.0	353.1	77.4	72.8	503.3	34.07	30.2
16802	Tail	565.5			104.6	420.6	27.6	6.0	2.3	35.9	3.11	1.9
	Total for meats		63,087.3	86.78	25,178.8	30,486.2	4,930.7	646.0	554.0	6,130.7	205.52	351.4
16804	Bones (less marrow)		5,394.0	7.42	2,292.5	759.5	988.1	46.9	50.7	1,085.7	10.79	1,364.7
16805	Marrow		92.9	0.13	13.3	75.2	1.8	0.17	0.03	2.0	0.06	
16807	Skin		3,857.0	5.30	a1,787.0	a882.5	a504.9	a334.4	a177.0	a1,016.3	a7.33	a23.9
16809	Spinal cord		06.0	0.09	a31.9	a27.2	a0.6	a0.6	a0.2	a4.5	b0.05	a0.3
16811	Tendons		133.5	0.18	a78.8	a14.9	a32.0	a5.8	a1.5	a39.3	a0.25	a1.4
17184	Hoofs		75.0	0.10	39.2	0.4				35.1		0.5
	Total weights		72,705.7	100.00	29,361.5	32,246.5	6,461.2	1,033.87	783.43	8,313.6	224.00	1,742.2
	Total per cents of original material				40.39	44.35	8.89	1.42	1.08	11.44	0.31	2.40

a Calculated from averages of like cuts.   *b* In residue and fat extract.

TABLE 8 A.—*Chemical composition of the meat of the pigs, by cuts.*

AMERICAN CLEAR BACKS.

[Data are stated in percentages of the original material.]

Serial No.	Pig, number and variety.	Water.	Fat.	Nitrogenous substances.				Leci-thin. a	Ash.	Total.
				Pro-teids in-soluble in hot water.	Gelati-noids.	Flesh bases.	Total.			
16667	1. Berkshire	32.27	57.09	7.00	0.50	0.91	8.41	0.15	0.51	99.03
16696	2. Tamworth	20.13	61.70	5.75	0.69	0.91	7.35	0.13	0.43	98.80
10609	3. Chester White	23.72	70.16	4.50	6.44	0.75	5.69	0.12	0.35	100.04
16638	4. Poland China	26.13	66.33	5.13	0.56	0.81	6.50	b 0.21	0.38	99.55
10579	5. Duroc Jersey	20.75	73.25	3.94	0.50	0.84	5.28	0.08	0.29	99.65
16725	6. Duroc Jersey	20.32	73.03	3.75	0.44	0.72	4.91	c 0.20	0.29	99.35
16754	7. Duroc Jersey	20.23	73.95	4.06	0.56	0.50	5.12	c 0.48	0.29	100.07
16783	8. Yorkshire	28.55	62.53	5.94	1.06	0.62	7.62	c 0.32	0.41	99.43
	Means	25.14	67.41	5.01	0.59	0.76	6.36	0.21	0.37	99.49
	Maxima	32.27	73.95	7.00	1.06	0.91	8.41	0.48	0.51	100.07
	Minima	20.23	57.69	3.75	0.44	0.50	4.91	0.08	0.29	98.80

TABLE 8 B.—*Chemical composition of the meat of the pigs, by cuts.*

AMERICAN CLEAR BELLIES.

[Data are stated in percentages of the original material.]

Serial No.	Pig, number and variety.	Water.	Fat.	Nitrogenous substances.				Leci-thin.a	Ash.	Total.
				Pro-teids in-soluble in hot water.	Gelati-noids.	Flesh bases.	Total.			
16669	1. Berkshire	37.27	51.93	7.00	0.56	1.22	8.78	0.14	0.55	98.67
16698	2. Tamworth	33.69	56.52	7.00	0.63	0.91	8.54	0.15	0.47	99.37
10611	3. Chester White	30.54	60.73	5.44	0.63	1.03	7.10	0.08	0.42	98.87
16640	4. Poland China	30.78	60.69	4.81	0.63	1.19	6.63	0.12	0.43	98.65
16581	5. Duroc Jersey	29.13	62.83	5.06	0.50	1.00	6.62	0.10	0.42	99.10
16727	6. Duroc Jersey	34.52	58.97	4.38	0.56	0.59	5.53	c 0.12	0.22	99.36
16756	7. Duroc Jersey	21.53	73.56	2.81	0.38	0.21	3.50	c 0.48	0.21	99.28
16785	8. Yorkshire	33.79	56.22	6.75	1.25	0.66	8.66	c 0.25	0.48	99.40
	Means	31.41	60.18	5.41	0.65	0.86	6.92	0.18	0.48	99.09
	Maxima	37.27	73.56	7.00	1.25	1.22	8.78	c 0.48	0.55	99.40
	Minima	21.53	51.93	2.81	0.38	0.31	3.50	0.08	0.21	98.65

TABLE 8 C.—*Chemical composition of the meat of the pigs, by cuts.*

SHORT CUT HAMS.

[Data are stated in percentages of the original material.]

Serial No.	Pig, number and variety.	Water.	Fat.	Nitrogenous substances.				Leci-thin.a	Ash.	Total.
				Pro-teids in-soluble in hot water.	Gelati-noids.	Flesh bases.	Total.			
10671	1. Berkshire	60.29	22.19	14.00	0.69	1.15	15.84	c 0.05	0.96	99.93
16700	2. Tamworth	57.93	24.45	13.00	0.63	1.25	14.88	0.22	0.84	98.32
16613	3. Chester White	53.15	30.99	11.13	0.63	1.97	13.73	0.35	0.80	99.02
16642	4. Poland China	54.78	30.12	10.69	0.81	1.28	12.78	0.23	0.76	98.67
16583	5. Duroc Jersey	50.45	35.94	8.38	1.00	0.50	9.88	0.03	0.71	97.01
16729	6. Duroc Jersey	37.26	39.10	17.38	1.25	1.69	20.32	0.35	1.32	98.35
10758	7. Duroc Jersey	44.26	43.38	9.56	0.87	0.81	11.24	c 0.45	0.65	99.98
16787	8. Yorkshire	59.14	25.23	11.50	0.87	1.09	13.46	c 0.42	0.74	98.99
	Means	52.16	30.18	11.96	0.84	1.22	14.02	0.24	0.85	98.79
	Maxima	60.29	43.38	17.38	1.25	1.97	20.32	c 0.65	1.32	99.98
	Minima	37.26	22.19	8.38	0.63	0.81	9.88	0.03	0.65	97.01

a In extracted sample.   *b* Calculated from averages of like cuts.   *c* In residue and fat extract.

TABLE 8 D.—*Chemical composition of the meat of the pigs, by cuts.*

NEW YORK SHOULDERS.

[Data are stated in percentages of the original material.]

Serial No.	Pig, number and variety.	Water.	Fat.	Proteids insoluble in hot water.	Gelatinoids.	Flesh bases.	Total.	Lecithin. a	Ash.	Total.
				Nitrogenous substances.						
16673	1. Berkshire............	54.97	29.01	11.25	0.81	1.56	13.62	0.15	0.89	98.64
16702	2. Tamworth	55.07	29.98	11.25	0.87	1.19	13.31	0.24	0.79	99.30
16615	3. Chester White.....	49.16	37.62	9.44	0.87	1.34	11.65	0.28	0.71	99.42
16044	4. Poland China	51.72	33.74	7.81	0.63	0.94	9.38	0.30	0.59	95.73
16585	5. Duroc Jersey	44.16	43.74	7.03	0.63	1.00	9.35	0.10	0.55	97.90
16731	6. Duroc Jersey	39.46	49.18	7.50	0.94	1.06	9.50	0.12	0.84	99.10
16760	7. Duroc Jersey	44.62	42.65	9.06	0.94	0.87	10.87	b 0.62	0.91	99.67
16789	8. Yorkshire	49.57	36.09	8.50	0.87	1.09	10.46	0.17	0.63	90.92
	Means............	48.59	37.75	9.06	0.82	1.14	11.02	0.25	0.74	98.35
	Maxima	55.07	49.18	11.25	0.94	1.56	13.62	b 0.62	0.91	99.67
	Minima.........	39.46	29.01	7.50	0.63	0.87	9.35	0.10	0.55	95.73

TABLE 8 E.—*Chemical composition of the meat of the pigs, by cuts.*

FEET.

[Data are stated in percentages of the original material.]

Serial No.	Pig, number and variety.	Water.	Fat.	Proteids insoluble in hot water.	Gelatinoids.	Flesh bases.	Total.	Lecithin. a	Ash.	Total.
				Nitrogenous substances.						
16675	1. Berkshire............	61.28	16.83	12.19	4.69	2.34	19.22	b 0.61	0.82	98.76
16704	2. Tamworth	58.66	21.23	11.63	3.38	2.00	17.97	b 0.51	0.86	99.23
16617	3. Chester White.....	53.05	26.74	9.88	6.38	1.06	17.32	0.10	0.84	98.04
16646	4. Poland China	50.60	31.32	11.00	4.19	0.97	16.16	0.15	0.91	99.20
16587	5. Duroc Jersey	54.16	26.19	10.19	2.10	2.90	15.28	c 0.32	0.76	96.71
16733	6. Duroc Jersey	51.84	29.47	10.50	2.88	2.31	15.75	b 0.35	0.75	98.10
16762	7. Duroc Jersey	55.98	25.25	12.38	2.63	1.53	16.54	b 0.36	0.78	98.91
16791	8. Yorkshire..........	57.47	30.86	5.13	1.63	1.53	8.29	0.08	0.41	97.11
	Means...........	55.39	25.99	10.37	3.50	1.05	15.82	0.32	0.77	98.28
	Maxima	61.28	31.32	12.38	6.38	2.96	19.22	b 0.61	0.91	99.23
	Minima..........	50.66	16.83	5.13	1.63	0.97	8.29	0.08	0.41	96.71

TABLE 8 F.—*Chemical composition of the meat of the pigs, by cuts.*

SPARERIBS.

[Data are stated in percentages of the original material.]

Serial No.	Pig, number and variety.	Water.	Fat.	Proteids insoluble in hot water.	Gelatinoids.	Flesh bases.	Total.	Lecithin. a	Ash.	Total.
				Nitrogenous substances.						
16677	1. Berkshire............	52.54	29.10	13.44	1.13	1.19	15.76	0.35	1.00	98.75
16706	2. Tamworth	49.20	33.88	11.56	1.31	1.40	14.27	0.25	0.93	98.53
16619	3. Chester White.....	53.23	27.93	13.63	0.87	1.65	16.15	0.28	0.92	98.51
16648	4. Poland China	52.95	29.55	10.06	0.63	3.09	13.78	0.31	0.95	97.54
16589	5. Duroc Jersey	54.09	26.90	14.06	0.81	1.53	16.40	0.12	1.04	98.55
16735	6. Duroc Jersey	49.84	31.95	13.63	1.13	1.09	15.85	0.33	1.10	99.07
16764	7. Duroc Jersey	53.20	27.51	14.56	1.13	1.09	16.78	b 0.83	1.01	99.33
16793	8. Yorkshire	52.31	29.28	13.56	1.13	1.34	16.03	0.33	1.05	99.00
	Means............	52.17	29.51	13.06	1.02	1.55	15.63	0.35	1.00	98.66
	Maxima	54.00	33.88	14.56	1.31	3.09	16.78	b 0.83	1.10	99.33
	Minima..........	49.20	26.90	10.06	0.63	1.00	13.78	0.12	0.92	97.54

a In extracted sample.   b In residue and fat extract.   c Calculated from averages of like cuts.

TABLE 8 G.—*Chemical composition of the meat of the pigs, by cuts.*

TENDERLOINS.

[Data are stated in percentages of the original material.]

Serial No.	Pig, number and variety.	Water.	Fat.	Proteids insoluble in hot water.	Nitrogenous substances.			Lecithin.a	Ash.	Total.
					Gelatinoids.	Flesh bases.	Total.			
16679	1. Berkshire	68.06	8.78	18.56	0.50	1.06	20.12	0.49	1.17	98.62
16708	2. Tamworth	65.52	13.51	17.13	0.56	0.91	18.60	b 0.91	1.06	99.60
16621	3. Chester White	65.97	13.47	16.69	0.44	1.03	18.16	0.40	1.06	99.06
16650	4. Poland China	67.43	10.95	17.63	0.69	0.78	19.10	0.39	1.13	99.00
16591	5. Duroc Jersey	66.55	11.41	18.31	0.38	1.09	19.78	0.35	1.14	99.23
16737	6. Duroc Jersey	63.86	14.81	17.31	0.63	0.87	18.81	0.56	1.05	99.09
16766	7. Duroc Jersey	72.83	9.35	14.69	0.50	0.56	15.75	0.32	0.86	99.11
16795	8. Yorkshire	66.65	13.07	16.00	0.81	1.03	17.84	b 0.68	1.04	99.28
	Means	67.11	11.92	17.04	0.56	0.92	18.52	0.51	1.06	99.12
	Maxima	72.83	14.81	18.56	0.81	1.09	20.12	b 0.91	1.17	99.60
	Minima	63.86	8.78	14.69	0.44	0.56	15.75	0.32	0.86	98.62

TABLE 8 H.—*Chemical composition of the meat of the pigs, by cuts.*

NECK BONES.

[Data are stated in percentages of the original material.]

Serial No.	Pig, number and variety.	Water.	Fat.	Proteids insoluble in hot water.	Nitrogenous substances.			Lecithin.a	Ash.	Total.
					Gelatinoids.	Flesh bases.	Total.			
16680	1. Berkshire	55.70	27.02	12.44	0.75	1.06	14.25	b 0.68	0.81	99.36
16709	2. Tamworth	55.52	26.03	13.38	1.31	1.40	16.09	b 0.83	1.02	99.49
16622	3. Chester White	53.50	29.52	12.19	0.75	1.31	14.25	0.25	0.87	98.39
16651	4. Poland China	55.23	28.27	10.25	1.13	1.75	13.13	b 0.45	0.91	97.99
16592	5. Duroc Jersey	52.48	30.31	12.50	0.75	1.40	14.65	0.20	0.94	98.58
16738	6. Duroc Jersey	49.30	34.92	11.60	0.94	1.06	13.69	0.28	0.87	99.06
16767	7. Duroc Jersey	52.10	30.82	12.38	1.13	0.62	14.13	b 0.55	0.94	98.64
16796	8. Yorkshire	55.13	26.71	13.31	1.00	1.15	15.46	b 0.54	0.99	98.83
	Means	53.62	29.33	12.27	0.97	1.22	14.46	0.47	0.92	98.79
	Maxima	55.70	34.92	13.38	1.31	1.75	16.09	b 0.83	1.02	99.49
	Minima	49.30	26.03	10.25	0.75	0.62	13.13	0.20	0.81	97.99

TABLE 8 I.—*Chemical composition of the meat of the pigs, by cuts.*

BACKBONES.

[Data are stated in percentages of the original material.]

Serial No.	Pig, number and variety.	Water.	Fat.	Proteids insoluble in hot water.	Nitrogenous substances.			Lecithin.a	Ash.	Total.
					Gelatinoids.	Flesh bases.	Total.			
16682	1. Berkshire	52.83	27.22	14.38	0.87	1.44	16.69	0.26	1.24	98.24
16711	2. Tamworth	51.06	30.67	13.69	0.87	1.22	15.78	0.25	1.10	98.86
16624	3. Chester White	50.39	31.65	13.38	0.56	1.34	15.28	0.23	1.05	98.60
16653	4. Poland China	51.26	30.98	12.94	0.81	1.34	15.09	0.27	1.05	98.65
16594	5. Duroc Jersey	48.77	29.40	16.38	0.81	1.65	18.84	0.41	1.23	98.65
16740	6. Duroc Jersey	47.54	35.96	12.81	1.06	1.15	15.02	0.36	0.92	99.80
16769	7. Duroc Jersey	53.09	28.07	13.88	1.06	1.19	16.13	b 0.44	1.02	98.75
16798	8. Yorkshire	50.65	29.61	14.13	1.25	1.28	16.66	b 0.78	1.18	98.88
	Means	50.70	30.45	13.95	0.91	1.33	16.19	0.38	1.11	94.81
	Maxima	53.09	35.96	16.38	1.25	1.65	18.84	b 0.78	1.24	99.80
	Minima	47.54	27.22	12.81	0.56	1.15	15.02	0.23	0.92	98.24

a In extracted sample.   b In residue and fat extract.

TABLE 8 J.—*Chemical composition of the meat of the pigs, by cuts.*

TRIMMINGS.

[Data are stated in percentages of the original material.]

Serial No.	Pig, number and variety.	Water.	Fat.	Nitrogenous substances.				Lecithin. a	Ash.	Total.
				Proteids insoluble in hot water.	Gelatinoids.	Flesh bases.	Total.			
16684	1. Berkshire	29.68	62.00	5.19	0.69	1.03	6.91	0.11	0.41	99.11
16713	2. Tamworth	28.85	62.67	5.31	0.81	1.03	7.15	0.10	0.43	99.20
16626	3. Chester White	22.49	71.557	3.25	0.50	0.84	4.59	0.07	0.31	99.01
16655	4. Poland China	22.43	71.52	3.66	0.44	0.69	4.82	0.06	0.32	99.15
16596	5. Duroc Jersey	20.08	74.13	3.25	0.44	0.87	4.56	0.08	0.30	99.15
10742	6. Duroc Jersey	16.51	78.78	3.06	0.38	0.50	3.94	0.07	0.25	99.54
16771	7. Duroc Jersey	20.49	73.56	3.38	0.63	0.75	4.76	b 0.11	0.28	99.20
16800	8. Yorkshire	26.12	65.81	4.56	1.00	0.94	6.50	b 0.44	0.39	99.26
	Means	23.33	70.00	3.96	0.61	0.83	5.40	0.13	0.34	99.20
	Maxima	29.68	78.78	5.31	1.00	1.03	7.15	b 0.44	0.43	99.54
	Minima	16.51	62.00	3.06	0.38	0.50	3.94	0.06	0.25	99.01

TABLE 8 K.—*Chemical composition of the meat of the pigs, by cuts.*

TAIL.

[Data are stated in percentages of the original material.]

Serial No.	Pig, number and variety.	Water.	Fat.	Nitrogenous substances.				Lecithin. a	Ash.	Total.
				Proteids insoluble in hot water.	Gelatinoids.	Flesh bases.	Total.			
16686	1. Berkshire	24.02	68.23	5.75	0.56	0.50	6.81	0.17	0.39	99.62
16715	2. Tamworth	25.77	67.08	4.25	0.69	0.78	5.72	0.10	0.30	98.97
16628	3. Chester White	15.66	79.69	3.00	0.44	0.62	4.06	0.08	0.24	99.73
16657	4. Poland China	16.50	77.77	3.44	0.56	0.66	4.66	0.09	0.31	99.33
16598	5. Duroc Jersey	11.54	84.62	2.00	0.31	0.62	2.93	0.07	0.20	99.36
16744	6. Duroc Jersey	13.94	81.23	3.00	0.44	0.41	3.85	0.06	0.23	99.31
16773	7. Duroc Jersey	13.73	81.74	2.81	0.50	0.44	3.75	b 0.48	0.27	99.97
16802	8. Yorkshire	18.50	74.37	4.88	1.06	0.41	6.35	b 0.55	0.33	100.10
	Means	17.46	76.84	3.64	0.57	0.56	4.77	0.20	0.28	99.55
	Maxima	25.77	84.62	5.75	1.06	0.78	6.81	b 0.55	0.39	100.10
	Minima	11.54	67.08	2.00	0.31	0.41	2.93	0.06	0.20	99.31

TABLE 9.—*Average composition of the meats from all the cuts of each animal.*

[Percentages.]

Number and name of pig.	Water.	Fat.	Nitrogenous substances.				Lecithin. c	Ash.	Total.
			Proteids insoluble in hot water.	Gelatinoids.	Flesh bases.	Total.			
1. Berkshire	43.12	43.98	9.21	0.67	1.15	11.03	0.25	0.68	99.06
2. Tamworth	40.60	47.22	8.52	0.76	1.07	10.35	0.17	0.60	98.94
3. Chester White	35.13	54.83	6.84	0.63	1.13	8.60	0.18	0.52	99.26
4. Poland China	37.17	52.27	6.65	0.65	1.03	8.33	0.16	0.52	98.45
5. Duroc Jersey	31.82	59.08	5.81	0.62	0.89	7.32	0.09	0.46	98.67
6. Duroc Jersey	29.42	60.53	6.95	0.70	0.88	8.53	0.18	0.54	90.20
7. Duroc Jersey	29.63	61.83	5.82	0.67	0.64	7.13	0.44	0.46	99.49
8. Yorkshire	30.91	48.32	7.82	1.02	0.88	9.72	0.33	0.56	98.84
Means	35.85	53.51	7.20	0.72	0.96	8.88	0.23	0.54	98.99
Maxima	43.12	61.83	9.21	1.02	1.15	11.03	0.44	0.68	99.26
Minima	29.42	43.98	5.81	0.62	0.64	7.13	0.09	0.46	98.45

a In extracted sample.   b In residue and fat extract.
c In the residue after the removal of the fat.

TABLE 10.—*Averages computed from all the bones of each cut of each animal, without marrow.*

[Percentages.]

Number and name of pig.	Water.	Fat.	Nitrogenous substances.				Lecithin. a	Ash.	Total.
			Proteids insoluble in hot water.	Gelatinoids.	Flesh bases.	Total.			
1. Berkshire	38.94	11.67	17.50	0.38	1.25	19.13	0.44	26.12	96.30
2. Tamworth	38.06	15.06	17.56	0.09	0.81	19.06	0.04	25.35	97.57
3. Chester White	40.41	17.18	17.57	0.50	1.03	19.10	0.06	21.69	98.44
4. Poland China	42.70	9.87	16.51	1.75	1.56	19.82	0.53	25.25	98.17
5. Duroc Jersey	36.45	12.89	20.13	0.56	1.12	21.81	0.50	26.66	98.31
6. Duroc Jersey	33.78	17.64	17.69	1.00	1.22	19.91	b 0.31	26.06	97.70
7. Duroc Jersey	36.64	13.70	18.94	0.56	1.15	20.65	c 0.43	27.07	98.49
8. Yorkshire	41.39	14.08	18.32	0.87	0.94	20.13	0.20	25.30	101.10
Means	38.55	14.01	18.03	0.79	1.14	19.95	0.31	25.44	98.26
Maxima	42.70	17.64	20.13	1.75	1.56	21.81	0.53	27.07	101.10
Minima	33.78	9.87	16.51	0.38	0.81	19.06	0.04	21.69	96.30

a In the residue after the removal of the fat.
b Calculated from averages of like cuts.
c In residue and fat extract.

TABLE 11.—*Analytical data for marrow.*

[Percentages.]

Number and name of pig.	Water.	Fat.	Nitrogenous substances.				Lecithin. a	Ash.	Total.
			Proteids insoluble in hot water.	Gelatinoids.	Flesh bases.	Total.			
1. Berkshire	14.36	81.51	2.00	0.19	0.06	2.25	0.46	....	98.58
2. Tamworth	13.31	84.48	1.38	0.06	0.06	1.50	0.05	....	99.34
3. Chester White	15.50	79.86	1.88	0.06	0.06	2.00	(0.06)	....	97.42
4. Poland China	16.74	78.35	2.56	0.13	0.03	2.72	(0.06)	....	97.87
5. Duroc Jersey	13.22	80.97	2.88	0.19	0.09	3.16	(0.06)	....	97.41
8. Yorkshire	14.29	81.58	1.88	0.19	0.03	2.10	0.06	....	98.03
Means	14.57	81.13	2.10	0.14	0.06	2.29	0.13	....	98.11
Maxima	16.74	84.48	2.88	0.19	0.09	3.16	0.46	....	99.34
Minima	13.22	78.35	1.38	0.06	0.03	1.50	0.05	....	97.41

a In the residue after removal of the fat.

TABLE 12.—*Analytical data for skin.*

[Percentages.]

Number and name of pig.	Water.	Fat.	Nitrogenous substances.				Lecithin. a	Ash.	Total.
			Proteids insoluble in hot water.	Gelatinoids.	Flesh bases.	Total.			
1. Berkshire	50.24	17.11	25.25	6.69	1.37	33.31	b 0.41	0.63	101.70
2. Tamworth	55.38	14.15	16.62	9.00	3.12	28.74	b 0.25	0.85	99.17
3. Chester White	40.78	31.17	4.50	6.06	7.89	18.45	0.10	0.53	91.03
4. Poland China	46.81	23.31	8.87	11.44	5.37	25.68	0.25	0.63	96.68
5. Duroc Jersey	35.49	38.16	3.69	11.38	4.93	20.00	0.09	0.48	94.22
6. Duroc Jersey	45.20	20.59	13.50	6.00	6.08	25.58	0.06	0.63	92.06
7. Duroc Jersey	50.39	15.75	19.19	10.13	3.34	32.66	b 0.15	0.78	99.73
Means	46.33	22.89	13.09	8.67	4.59	26.35	0.19	0.62	96.52
Maxima	55.38	38.16	25.25	11.44	7.89	33.31	0.41	0.78	101.70
Minima	35.49	14.15	3.69	6.00	1.37	18.45	0.06	0.48	91.03

a In the residue, after removal of the fat.   b In residue and fat extract.

TABLE 13.—*Analytical data for spinal cord.*

[Percentages.]

Number and name of pig.	Water.	Fat.	Nitrogenous substances.				Lecithin. a	Ash.	Total.
			Proteids insoluble in hot water.	Gelatinoids.	Flesh bases.	Total.			
1. Berkshire	65.70	20.76	3.88	0.69	0.16	4.73	b 1.47		98.66
2. Tamworth	46.45	45.39	4.06	0.63	0.34	5.03	c 2.95	0.23	100.05
4. Poland China	48.86	42.94	5.50	1.06	0.22	6.78	c 1.10	0.56	100.24
5. Duroc Jersey	59.50	23.62	8.56	0.94	0.47	9.97	b 1.47		94.56
7. Duroc Jersey	20.84	67.32	6.06	1.00	0.28	7.34	c 0.70		96.20
Means	48.27	41.21	5.61	0.86	0.29	6.77	c 1.54	0.39	97.94
Maxima	65.70	67.32	8.56	1.06	0.47	9.97	c 2.95	0.56	100.24
Minima	20.84	23.62	3.88	0.63	0.16	4.73	c 0.70	0.23	94.56

a In the residue after removal of the fat.   b In fat extract calculated from averages of like cuts.   c In fat extract.

TABLE 14.—*Analytical data for tendons.*

[Percentages.]

Number and name of pig.	Water.	Fat.	Nitrogenous substances.				Lecithin. a	Ash.	Total.
			Proteids insoluble in hot water.	Gelatinoids.	Flesh bases.	Total.			
1. Berkshire	58.43	13.40	22.24	4.44	0.62	27.50	b 0.45	1.18	100.96
2. Tamworth	61.55	8.09	24.19	4.31	1.37	29.87	0.10	1.09	100.70
3. Chester White	60.12	10.41	24.25	3.25	1.31	28.81	0.28	0.85	100.47
4. Poland China	56.68	11.68	24.75	5.50	0.97	31.22	0.10	1.48	101.16
6. Duroc Jersey	59.62	9.32	25.25	4.63	1.31	31.19	0.08	0.96	101.17
7. Duroc Jersey	57.91	14.02	22.94	3.94	0.97	27.85	0.11	0.87	100.76
Means	59.05	11.15	23.97	4.35	1.10	29.41	0.19	1.07	100.87
Maxima	61.55	14.02	25.25	5.50	1.37	31.22	b 0.45	1.48	101.17
Minima	56.68	8.09	22.44	3.25	0.62	27.50	0.08	0.85	100.47

a In the residue after the removal of the fat.   b In residue and fat extract.

TABLE 15.—*Analytical data for hoofs.*

[Percentages.]

Number and name of pig.	Water.	Fat.	Nitrogenous substances.				Ash.	Total.
			Proteids insoluble in hot water.	Gelatinoids.	Flesh bases.	Total.		
1. Berkshire	41.09	0.86				58.00	0.93	100.88
2. Tamworth	43.47	0.61				55.63	0.98	100.69
3. Chester White	39.31	0.70				60.25	0.89	101.15
4. Poland China	47.52	0.60				52.31	0.81	101.24
5. Duroc Jersey	35.10	0.74				64.63	0.85	101.32
6. Duroc Jersey	44.68	0.69				53.50	1.02	99.89
7. Duroc Jersey	47.12					55.69		
8 Yorkshire	52 26	0.49				46.75	0.71	100.21
Means	43.82	0.67				55.85	0.88	100.77
Maxima	52.26	0.86				64.63	1.02	101.32
Minima	35.10	0.49				46.75	0.71	99.89

TABLE 16.—*Weights of the entire animals and their various cuts, as weighed in Chicago and in Washington, together with the apparent percentages of gain or loss in transit.*

Number and name of pig.	Two clear backs.		Two clear bellies.		Two short-cut hams.		Two New York shoulders.	
	Chicago.	Washington.	Chicago.	Washington.	Chicago.	Washington.	Chicago.	Washington.
1. Berkshire:								
Pounds	35½	34⅜	19½	19¼	23½	23 5/16	20½	20⅝
Grams	16,102.8	15,592.5	8,845.2	8,731.8	10,659.6	10,574.6	9,208.8	9,395.5
2. Tamworth:								
Pounds	41	40½	20	19 9/16	26	25¾	21	20¾
Grams	18,597.6	18,370.8	9,072.0	8,873.0	11,793.6	11,680.2	9,525.6	9,412.2
3. Chester White:								
Pounds	36	35¾	21	21	20	19 15/16	21	20 13/16
Grams	16,329.6	16,210.2	9,525.6	9,525.6	9,072.0	9,057.8	9,525.6	9,440.5
4. Poland China:								
Pounds	40	38¾	24	23⅝	26	25⅞	24	23 7/16
Grams	18,144.0	17,577.0	10,886.2	10,716.1	11,793.6	11,736.9	10,886.2	10,631.1
5. Duroc Jersey:								
Pounds	39½	39½	24	24⅛	21	21	19½	19 13/16
Grams	17,917.2	17,917.2	10,886.4	10,943.1	9,525.6	9,525.6	8,845.2	8,967.0
6. Duroc Jersey:								
Pounds	45	44½	32½	32 11/16	27	26⅞	22	22⅛
Grams	20,412.0	20,185.2	14,742.0	14,827.1	12,247.2	12,190.5	9,979.2	10,035.9
7. Duroc Jersey:								
Pounds	39½	38⅞	28½	29	23½	23 13/16	19½	19¾
Grams	17,917.2	17,633.7	12,927.6	13,154.4	10,659.6	10,801.4	8,845.2	8,958.6
8. Yorkshire:								
Pounds	44	43 13/16	22½	23 1/16	27	27¼	24½	25 9/16
Grams	19,958.4	19,873.4	10,206.0	10,461.2	12,247.2	12,360.6	11,113.2	12,502.4

Number and name of pig.	Four feet.		Spareribs.		Tenderloins.		Neck bones.	
	Chicago.	Washington.	Chicago.	Washington.	Chicago.	Washington.	Chicago.	Washington.
1. Berkshire:								
Pounds	3½	3⅓	5	4⅞	1	1⅛	2	1⅞
Grams	1,594.2	1,514.1	2,268.0	2,212.0	453.6	470.8	907.2	842.5
2. Tamworth:								
Pounds	4½	4⅓	5	4 7/10	1	1⅙	2	1 9/10
Grams	2,057.3	1,974.1	2,268.0	2,132.7	453.6	528.2	907.2	886.0
3. Chester White:								
Pounds	2½	2 7/10	3	3 1/10	1	1	1½	1 9/16
Grams	1,152.5	1,236.9	1,360.8	1,409.0	453.6	453.6	680.4	683.7
4. Poland China:								
Pounds	3	2 15/16	5	4⅓	1	9/10	1½	1⅘
Grams	1,360.8	1,359.0	2,268.0	1,969.5	453.6	419.8	680.4	815.6
5. Duroc Jersey:								
Pounds	2½	2¾	3½	3⅝	1	¾	1½	1¾
Grams	1,137.9	1,255.4	1,587.6	1,612.0	453.6	348.5	680.4	784.0
6. Duroc Jersey:								
Pounds	3½	3⅜	4	3 9/16	½	9/10	2	1 9/10
Grams	1,587.6	1,547.2	1,814.4	1,612.2	226.8	421.3	907.2	892.6
7. Duroc Jersey:								
Pounds	2½	3 1/16	3½	3⅓	¾	¾	2	1 15/16
Grams	1,134.0	1,400.0	1,587.6	1,504.0	340.2	333.3	907.2	906.5
8. Yorkshire:								
Pounds	4½	4 15/16	5	5⅛	1	1⅜	2	2⅝
Grams	2,041.2	2,246.0	2,268.0	2,340.0	453.6	632.5	907.2	1,192.3

Number and name of pig.	Backbones.		Trimmings.		Tail.		Total.		Per cent of gain or loss in transit.
	Chicago.	Washington.	Chicago.	Washington.	Chicago.	Washington.	Chicago.	Washington.	
1. Berkshire:									
Pounds	3½	3 7/16	18	16 9/16	¼	⅘	132¼	129⅗	Loss.
Grams	1,587.6	1,580.0	8,164.8	7,512.8	113.4	363.0	59,905.2	58,789.6	2.01
2. Tamworth:									
Pounds	4	4 1/16	18¼	16⅝	¼	1 9/16	143	140 9/10	Loss.
Grams	1,814.4	1,840.0	8,278.2	7,541.1	113.4	707.5	64,880.9	63,946.4	1.44
3. Chester White:									
Pounds	2½	2 9/16	27	15¾	¼	1⅝	135¾	125⅞	Loss.
Grams	1,134.0	1,172.6	12,247.2	7,144.2	113.4	740.2	61,594.7	57,080.3	8.07

TABLE 16.—*Weights of the entire animals and their various cuts, etc.*—Continued.

Number and name of pig.	Backbones.		Trimmings.		Tail.		Total.		Per cent of gain or loss in transit.
	Chicago.	Washington.	Chicago.	Washington.	Chicago.	Washington.	Chicago.	Washington.	
4. Poland China:									
Pounds	3	2$\frac{1}{16}$	21$\frac{1}{4}$	20	$\frac{1}{4}$	1$\frac{7}{16}$	149	140$\frac{1}{4}$	Loss.
Grams	1,360.8	1,315.5	9,639.0	9,072.0	113.4	760.0	67,586.4	66,372.5	1.80
5. Duroc Jersey:									
Pounds	3	3$\frac{1}{6}$	20$\frac{1}{4}$	19$\frac{11}{16}$	$\frac{1}{4}$	1$\frac{9}{16}$	136	138$\frac{1}{8}$	Gain.
Grams	1,360.8	1,438.0	9,185.4	8,930.3	113.4	683.0	61,093.5	62,424.1	1.18
6. Duroc Jersey:									
Pounds	3$\frac{1}{2}$	3$\frac{7}{8}$	27$\frac{1}{4}$	25$\frac{1}{8}$	$\frac{1}{4}$	2$\frac{7}{8}$	167$\frac{1}{2}$	167$\frac{1}{10}$	Loss.
Grams	1,587.6	1,546.0	12,360.6	11,309.4	113.4	1,173.0	75,978.0	75,799.4	0.27
7. Duroc Jersey:									
Pounds	3	3$\frac{1}{4}$	24$\frac{7}{8}$	23$\frac{11}{16}$	$\frac{1}{4}$	1$\frac{7}{16}$	147$\frac{3}{4}$	149$\frac{1}{4}$	Gain.
Grams	1,360.8	1,482.0	11,226.0	10,744.7	113.4	750.0	67,019.4	67,677.6	0.98
8. Yorkshire:									
Pounds	4	4$\frac{1}{2}$	24$\frac{7}{8}$	18$\frac{5}{8}$	$\frac{1}{4}$	1$\frac{7}{8}$	159$\frac{1}{2}$	160$\frac{1}{16}$	Gain.
Grams	1,814.4	1,998.0	11,226.6	8,448.3	113.4	651.0	72,349.2	72,705.7	0.49

TABLE 17.—*Relative proportions of parts of pigs, expressed in percentages, of the entire dressed animal, the head, leaf lard, and kidneys having been removed.*

Number and name of pig.	Weight in pounds (Washington).	Percentages of parts.						
		Meat (fat and lean).	Bones, less marrow	Marrow.	Skin.	Spinal cord.	Tendons.	Hoofs.	Total.
1. Berkshire	129.6	88.10	7.44	0.12	3.80	0.09	0.27	0.09	100
2. Tamworth	141	86.50	8.18	0.21	4.71	0.09	0.21	0.10	100
3. Chester White	125.8	87.94	6.21	0.08	5.52	0.07	0.12	0.06	100
4. Poland China	146.4	90.67	5.30	0.11	3.63	0.08	0.14	0.07	100
5. Duroc Jersey	137.6	88.03	5.92	0.11	5.75	0.04	0.10	0.05	100
6. Duroc Jersey	167.1	90.93	4.70	0.10	4.00	0.07	0.12	0.08	100
7. Duroc Jersey	149.2	89.90	5.07	0.11	4.65	0.08	0.11	0.08	100
8. Yorkshire	160.3	86.79	7.41	0.13	5.30	0.09	0.18	0.10	100
Means	144.6	88.62	6.28	0.12	4.67	0.08	0.16	0.08	100
Maxima	167.1	90.93	8.18	0.21	5.75	0.09	0.27	0.10	100
Minima	125.8	86.79	4.70	0.08	3.63	0.04	0.10	0.05	100

TABLE 18.—*Analytical data, expressed in percentages, of the entire dressed animal, the head, leaf lard, and kidneys having been removed.*

Number and name of pig.	Weight in pounds.	Water.	Fat.	Nitrogenous substances.				Lecithin.a	Ash.	Total.
				Proteids insoluble in hot water.	Gelatinoids.	Flesh bases.	Total.			
1. Berkshire	129$\frac{3}{4}$	43.10	40.46	10.45	0.89	1.16	13.02	0.27	2.57	99.42
2. Tamworth	141	41.09	42.97	9.65	1.15	1.14	11.90	0.17	2.63	98.85
3. Chester White	125$\frac{1}{2}$	35.80	51.11	7.39	0.92	1.50	9.85	0.17	1.84	98.77
4. Poland China	146$\frac{2}{5}$	37.83	48.90	7.27	1.11	1.21	9.66	0.19	1.83	98.41
5. Duroc Jersey	137$\frac{3}{5}$	32.32	55.07	6.55	1.24	1.14	8.99	0.11	2.01	98.50
6. Duroc Jersey	167$\frac{1}{10}$	30.31	56.81	7.73	0.93	1.11	9.80	0.19	1.75	98.86
7. Duroc Jersey	149$\frac{1}{4}$	30.58	57.68	7.03	1.10	0.78	8.96	0.42	1.81	99.45
8. Yorkshire	160$\frac{3}{10}$	40.39	44.35	8.89	1.42	1.08	11.44	0.31	2.40	98.89
Means	144$\frac{5}{8}$	36.43	49.67	8.12	1.10	1.14	10.46	0.23	2.11	98.90
Maxima	167$\frac{1}{10}$	43.10	57.68	10.45	1.42	1.50	13.02	0.42	2.63	99.45
Minima	125$\frac{5}{8}$	30.31	40.46	6.55	0.89	0.78	8.96	0.11	1.81	98.41

a In extracted residue, except as noted in preceding tables.

DISCUSSION OF THE DATA.

Tables 1 to 6, inclusive, contain the original analytical data from which the subsequent data showing the details of the composition of the meat were computed. The character of the data in these tables is pretty fully explained in a previous part of this report. These tables are particularly valuable, because they are the records of the data as made at the time the observations were made, and therefore show the extent and nature of the analytical work more elaborately than would be indicated by the details of tabular data shown in subsequent tables, which were obtained from a careful analytical study of Tables 1 to 6. It is believed that with the explanation previously given the student will be able to understand thoroughly the nature of the tables mentioned.

In Table 7 are found the general data in parts by weight for all the different parts and cuts of each animal. The footings show the total weight, in grams, of each constituent of each animal, and the second horizontal column of footings shows the percentage by weight of each constituent for each animal. The data in Table 7 are calculated from the original data contained in Tables 1 to 6, inclusive. The captions of Table 7 will explain sufficiently the nature of the data.

COMPOSITION OF THE SAME CUTS FROM THE DIFFERENT ANIMALS.

Tables 8 A to 8 K, inclusive, contain a comparison of the composition of the meat of the same cuts of each animal. Each table in the caption designates the character of the cut of meat on which the comparison is made. For instance—

Clear backs.—Table 8 A is a comparison of the composition of the meat of the American clear backs of all the animals. A study of the data reveals quite a variation in the composition of the meat from the different animals, and this variation is found in all the series of data. As in the other cases, we find that there is a corresponding relationship between the water and fat, one varying inversely as the other, so that the sum of the two is almost a constant quantity. The extremes of variation in water are found in the Berkshire and Duroc Jersey, namely, 32.27 and 20.23 per cent, respectively. The extremes of fat are also found in the same animals, namely, 57.69 and 73.95, respectively. In nitrogenous substances, as would naturally be expected, there is a corresponding variation, the samples which have the most fat, as a rule, having a lower percentage of nitrogenous bodies, and vice versa. This rule is not of rigid application, but must be regarded only in a general sense. For instance, in Table 8 A the largest percentage of nitrogenous substance is found in the Berkshire, which also has the smallest percentage of fat, while the smallest percentage of nitrogenous matter is found in the Duroc Jersey, No. 6, which, with one slight exception, has also the largest quantity of fat. The distri-

3020—No. 53——5

bution of the nitrogenous substances in the meats of the American clear backs is found in the table, where they are divided into three classes, namely, the true proteids, insoluble in hot water; gelatinoids, which are of a true proteid character, but soluble in hot water, and of which gelatin is the type; and the flesh bases, which are soluble in hot water and are not precipitated by the action of bromin. The ash, as would be expected in animal products, entirely free of bone, is not very large in quantity. It consists chiefly of common salt and the phosphates of the alkali metals. The sum of the substances obtained on analysis shows that very little of the whole matter was unaccounted for, and, when the nature of the material on which the work was done is considered, it is seen that the summation is eminently satisfactory.

Clear bellies.—In Table 8 B we find a study of the comparison of the meat of American clear bellies exactly analogous to that which has been described for the American clear backs. As a rule it will be seen that the percentage of water in the clear bellies is higher and the percentage of fat lower than in the American clear backs. The general remarks already made in regard to the clear backs may be applied to this table without tiresome repetition. The relations between the nitrogenous substances and the water and fat and the ash are practically the same as for those just described, while the summation of the analyses also shows a satisfactory accounting for the materials which the chemist is furnished. It will be noticed that the flesh bases in the clear bellies are higher than in the clear backs. Data of this kind are of a practical nature as well as of a scientific value, in indicating what portion of the carcasses of animals could best be used, for instance, for the manufacture of extracts. A similar study applied to beef cattle would reveal data of unusual interest in this respect. Again, we find the largest percentage of water in the case of the Berkshire, and also the smallest percentage of fat, while the smallest percentage of water and the largest percentage of fat are found in the Duroc Jersey, No. 7, this showing a remarkable concordance between the character of the meats of the two cuts in the various animals.

Short-cut hams.—Table 8 C contains a comparison of the data of the meat of short-cut hams. In this cut of meat is found a smaller percentage of fat, a correspondingly large percentage of water, and, of course, in the increase of the muscular tissue, a very largely increased amount of nitrogenous matters. Again, the largest quantity of water and the smallest quantity of fat are found in the meat of the Berkshire, while the smallest quantity of water is found in the Duroc Jersey, No. 6, and the largest quantity of fat in the Duroc Jersey, No. 7. The general relation of water and fat is thus found to be the same in this cut as in the two preceding ones. In regard to the nitrogenous substances there is quite a remarkable variation. The largest percentage of nitrogenous bodies is found in Duroc Jersey, No. 6, while the smallest is found in the Duroc Jersey, No. 5. It seems rather strange that two animals of the same breed show such a remarkable discrepancy in composition. In this instance, however, there is a deficit of material amounting to almost 3 per cent unaccounted for; so that the analytical data do not have the value which they would have did the summation reach more nearly 100. In the short-cut hams there is found a considerable increase in the quantity both of gelatinoid proteids and flesh bases over the amounts in the cuts already described.

New York shoulders.—Table 8 D contains comparisons of the meat of the cuts known as New York shoulders. In this cut we have a larger percentage of fat than in the one just described, and a correspondingly smaller quantity of water and a smaller quantity of nitrogenous bodies. The summation of the analyses is not as satisfactory as in most of the preceding cases, and in one case a deficit of 4¼ per cent is noticed. Working, however, with wet material, and in the manner which was made necessary in such an investigation, it is not to be wondered at that often discrepancies of this nature may occur. These discrepancies are probably due chiefly to the determinations of water and fat, which are the most difficult of all connected with the operation of determining the composition of fresh meats, and inasmuch as the water and fat constitute by far the largest portion of the material it is seen that these difficulties must now and then result in failing to secure in the summation an accounting for all the material present. The largest percentage of water in these cuts is found in the Tamworth, and the smallest percentage of fat in the Berkshire. The smallest percentage of water is found in Duroc Jersey, No. 6, and the largest percentage of fat in the same animal. The relation between the nitrogenous substances is sufficiently indicated in the table, and calls for no especial comment.

Feet.—Table 8 E contains a comparison of the composition of the meat of the feet of the different animals. In the feet we find a marked difference in the analytical data, and especially on account of the fact that the feet, as is well known, contain large quantities of gelatin, and, as the data show, also considerable quantities of flesh bases. The total quantities of nitrogenous matters, in proportion to the other materials, is much larger in the feet than in the preceding cuts, while the quantity of gelatin is shown with sufficient emphasis in the tables of analytical data. However, a remarkable variation from the type is found in the feet of the Yorkshire pig, where the total amount of nitrogenous matter is only about half of that of the other animals. The summation of this analysis shows approximately 100 per cent, and therefore the feet of this animal must be regarded as differing essentially from those of other pigs examined. In regard to the gelatin we find that the largest percentage is found in the feet of the Chester White, and the smallest in those of the Yorkshire. The largest quantity of nitrogenous matter is found in the feet of the Berkshire, and the smallest in the feet of the Yorkshire pig. Again, the Berkshire leads all the others in having a maximum quantity of water and a minimum quantity of fat in its feet. The smallest quantity of water was found in the feet of the Poland China, and the smallest quantity of fat in the feet of the Berkshire.

Spareribs.—Table 8 F contains a comparison of the composition of the meat of the spareribs. In this case the largest percentage of water was found in Duroc Jersey, No. 5, and the smallest in the Tamworth. The smallest quantity of fat was found in Duroc Jersey, No. 5, and the largest in the Tamworth. The spareribs are rich in nitrogenous matters, mostly of a proteid nature. The content of flesh bases in the Poland China is remarkably high, being nearly double that of the average. The summations of the analyses for this table are satisfactory.

Tenderloins.—Table 8 G contains a comparison of the tenderloins of the different animals. The maximum content of water in these cuts was found in Duroc Jersey, No. 7, and the minimum in Duroc Jersey, No. 6. The maximum content of fat is found in the Duroc Jersey, No. 6, and the minimum in the Berkshire. The tenderloins differ from all the preceding cuts in having a largely increased quantity of water and a decreased quantity of fat. On account of the muscular nature of the tissue the proportion of nitrogenous substances is larger than in any of the cuts preceding. These substances are mostly of a proteid nature, there being only a comparatively small quantity of gelatinoids and flesh bases. The ash of these meats is also quite high, showing a large content of mineral nutritive substances. The summations of the analyses are quite satisfactory.

Neck bones.—Table 8 H contains a comparison of the meat from the neck bones of the animal. These meats show quite a uniform composition, there being less variation among the different animals than in almost any of the cuts secured. For instance, the maximum content of water in these meats is 55.70 and the minimum 49.30, while the maximum content of fat is 34.92 and the minimum 26.03. There is also a quite uniform agreement in the content of nitrogenous substances as a whole and in each particular class, the variations being only nominal. The ash is also uniform in amount and the summation of the analyses satisfactory. The meat from the neck bones, therefore, shows the most uniform agreement in composition of different animals of any of the cuts yet studied.

Backbones.—Table 8 I contains a comparison of the composition of the meat from the backbones. There is also here a quite uniform agreement in the content of water and fat, the maximum content of water being 53.09 and the minimum 47.54, while in the case of the fat the maximum content is 35.96 and the minimum 27.22. The whole of the nitrogenous substances show also a greater uniformity, the only variation being in the case of Duroc Jersey, No. 5, where the total of the nitrogenous bodies is considerably higher than the mean of the other animals. Most of the nitrogenous matter in the meat of the backbones is protein, although the quantity of flesh bases is in every case more than 1 per cent. The ash is also quite high, showing a large proportion of nutritive mineral matters. The summation of the analyses is satisfactory.

Trimmings.—Table 8 J shows the composition of the trimmings from the different animals. These trimmings, as will be seen, consist chiefly of the fatty portions which are rejected in preparing the cuts for market. They are used principally for the manufacture of lard. They therefore show an excessively high content of fat and a comparatively low content of water and of nitrogenous bodies and ash. The summation of the analyses of these materials is therefore eminently satisfactory. The analytical data show that the trimmings from the different animals are quite uniform in composition.

Tails.—Table 8 K shows the composition of the meat cut from the tails of the animals. Here also we see a large excess of fat, a correspondingly small proportion of water and of nitrogenous bodies and of ash. The tail meats are not very concordant in their composition, there being large extremes shown in the proportions of the various constituents. This is in a large measure due to the carelessness of the cutters, as in some cases large quantities of fatty tissue were left connected with the cut designated as "tail," while in other cases the same portions of the animals were placed with the "trimmings." The largest amount of water in the tail meats is in the Tamworth, and the smallest in the Duroc Jersey, No. 5. The largest quantity of fat is found in the Duroc Jersey, No. 5, and the smallest in the Tamworth. The summation of the analyses here is also very satisfactory.

Average of all cuts.—Table 9 contains the average analyses of the meats of all of the cuts from each of the animals. These analyses were calculated from the preceding data, combining all of the meats into one expression for each animal. These data are true averages; that is, each part making up the mean in each case was given a weight according to the actual amount of matter which it represented. The data therefore show in a condensed form the variations between the composition of the meats of the different animals. It would not be fair to ascribe the differences which are noticed in the composition of the meats solely to the influence of the breed, because with the exception of one instance, where there are three animals of one breed, each breed is represented only by a single animal. In the case mentioned, however, where there are three animals representing the Duroc Jersey, it is seen that there is a marked agreement in the meat from each one. It is, therefore, fair to presume that the single animal for the other breeds represents fairly well types of that breed. With this statement the data have a greater value as showing the comparison between the meat of breeds than they would have had had there been only a single Duroc Jersey in the list. A study of the data shows that the Berkshire pig leads all others in having the maximum percentage of water and the minimum percentage of fat. The Berkshire, therefore, pound for pound, represents the least nutritive value of any of the breeds examined. Notwithstanding this fact, the Berkshire heads the list of all in its percentage of nitrogenous substances, and this compensates in a large degree for its increased percentage of water. There is quite a satisfactory agreement between the nitrogenous substances in the distribution thereof in the three classes named. The percentage of gelatinoid nitrogenous matters is fairly constant, only in one instance, namely, that of the Yorkshire, rising much above the average. All the other percentages are very near that of the mean.

In regard to the flesh bases, only one falls considerably below the average, namely, the Duroc Jersey, No. 7, the others being very close to the mean. In total nitrogen there is a marked deficit in the case of the Duroc Jersey, No. 7, but this is due not to the influence of breed alone upon the composition, but to the large excess of fat in the meat of this animal.

The ash shows a fairly constant number throughout, varying very little from the mean.

The summation of the analyses is fairly satisfactory. In no case is there as much as 2 per cent unaccounted for, the largest deficit being in the case of the Poland China, where it amounts to 1.55 per cent. When the nature of the material upon which the work was done is considered, the figures are eminently satisfactory. These data afford, it is believed, a better basis for nutritive studies of the meats of pigs than has heretofore been supplied from any chemical laboratory.

Average of bones.—Table 10 contains the average composition of all the bones of each animal. No separate analyses of the bones from each cut were made. For each pig one composite sample was made, including all the bones of the animal. As is to be expected in a case of this kind, it was found that the composition of the bones is reasonably uniform in the different animals. In regard to water, the largest quantity was found in the bones of the Poland China, namely, 42.70, and the smallest in the bones of the Duroc Jersey, No. 6, namely, 33.78 per cent. In regard to the content of fat, the largest quantity was found in the bones of the Duroc Jersey, No. 6, namely, 17.64 per cent, and the smallest in the bones of the Poland China, namely, 9.87 per cent. The bones are extremely rich in nitrogenous substances, and these consist mostly of the proteid matter insoluble in hot water. The quantity of gelatinous matter in bones is not so great as would be expected, being but little more, as a rule, than in the meats. On the other hand, the quantity of flesh bases is larger than would be expected, being considerably in excess of the quantity of gelatinous matter. The total quantity of nitrogenous matter in the different animals is remarkably near the mean, the mean quantity being 19.95 per cent and the variation not being quite 2 per cent in any case from the mean. The ash, naturally, is very high. The summation of the analyses is not as uniform as could be wished, ranging from 100.90 per cent as the maximum to 95.86 per cent as the minimum, a difference of little over 5 per cent. The difficulty of comminuting the bones into a homogeneous mass, and thus securing an average sample, probably accounts for a great deal of the discrepancy seen in the summations of the analyses. It is evident that the bones contain a very large amount of nutrient matter which would be available for digestion if they were sufficiently comminuted, since the ash consists almost exclusively of tricalcium phosphate, which is insoluble, and thus would not interfere greatly with the process of digestion. The bones of animals, however, are so valuable for fertilizing purposes that they have not been used to any extent for feeding, except for poultry.

Average of marrow.—Table 11 contains the average analyses of the samples of marrow from all the bones from each cut of each animal, except in the case of Duroc Jerseys, Nos. 6 and 7, where the samples of marrow were destroyed by mice. On account of the small amount of material at our disposal, the ash in the samples was not determined. The summation, therefore, represents only partially the total ingredients, since it does not include the ash nor the lecithin, which are very important components of the marrow substance. The marrow, as will be seen by the data, is essentially a fat product, more than 95 per cent of the whole weight of the material being composed of fat and water, the mean percentage of fat in the whole sample being 81.13, and of water, 14.57. The nitrogenous constituents of the marrow, while being extremely important from a physiological point of view, have not much value from a nutritive point. They constitute only 2.29 per cent of the whole. There is a fairly good concordance seen in the composition of the marrow from the different animals. In point of fat, the greatest variations are found in case of the Tamworth, with a maximum percentage of fat, and the Poland China, with a minimum percentage, the difference being, in round numbers, 16 per cent. The variations in water are less marked, while in the total nitrogenous matters only one, namely, the Tamworth, falls far below the others in the percentage contained. The summation is as good as could be expected, considering the fact that ingredients of considerable magnitude are omitted.

Average of skin.—Table 12 contains the average analytical data for the skin of all of the cuts of each animal. All the skin from each animal was mixed together and carefully comminuted by passing several times through a meat chopper until a homogeneous mass was obtained. From this mass a suitable sample was taken, representing as nearly as possible the average composition of the whole. On this were performed the analytical operations from which the data represented in Table 12 were secured. The table contains the analytical data for all the animals except No. 8, the Yorkshire, of which the sample was lost. The most remarkable fact in connection with a general view of the data is that the skins have a high rank among the nitrogenous substances of the animal. The mean percentage of nitrogenous matters in the skin is 26.35, and as the skin consists of almost half its weight of water, it is seen that the dry skin would contain 50 per cent of its weight of nitrogenous materials. The next most important ingredient is of course the fat, of which the average is 22.89. In the nitrogenous substances the proteids comprise about half of the whole. Of the other half two-thirds belong to the gelatinoids and one-third to the flesh bases. The skin, therefore, is preeminently a gelatinous body. About one-half of the total quantity of nitrogenous substances it contains is soluble in hot water, and one-third of the half which is soluble is not precipitated by bromin. If the gelatinous matters of the skin could be easily separated, they would be the most valuable parts of the animal for the preparation of the flesh bases. Skins of animals, however, are usually more valuable for the manufacture of leather than for any other purposes.

To go a little more into the detail of the data representing the composition of the skin, we find that the skin which had the largest percentage of water belonged to the Tamworth pig, and the one with the smallest to the Duroc Jersey, No. 5. Of fat the largest amount was found in the Duroc Jersey, No. 5, thus showing again the general relation of the proportions of water and fat to which attention has already been called. The smallest percentage of fat was found also in the case of the skin of the Tamworth, where the percentage of water was largest. In regard to nitrogenous substances the most remarkable variations are seen. In the Berkshire, which contained the largest proportion of nitrogenous substances, the true proteids comprise by far the larger portion, followed by the gelatinoids, while the flesh bases form a very small percentage of the whole. On the other hand, in the skin of the Duroc Jersey, No. 5 the quantity of proteids is comparatively small, while both the gelatinoids and flesh bases are high. Whether this marked peculiarity in the composition of the skin is due to the influence of the breed or to accidental causes can not be stated. Probably, however, it is due to accidental causes; as, for instance, the Chester White and the Duroc Jersey, No. 5 show similar composition of skins, but this is quite different from the composition of the skin of Duroc Jerseys, Nos. 6 and 7. It is possible, further, that owing to the peculiar structure of the skin and the difficulty of securing a homogeneous mixture of it, portions of the skin from different cuts vary relatively in the sample which was taken for analysis. Thus, for instance, if a portion of the skin very rich in gelatinous matter and flesh bases should form an excessive portion of the whole sample taken for analysis, the effect would be the same as is seen in the data recorded. The summation of the analyses is generally satisfactory, yet in one case there is a deficit of 9 per cent, while in another there is an excess of 1.70 per cent. These variations are doubtless due to the difficulty of securing a homogeneous sample for analytical purposes. Another source of unreliability in the samples of skin is found in the difficulty of avoiding variations in the amount of the underlying fatty tissue included in the sample. It is practically impossible to remove all of the tissue properly belonging with the skin without including a small quantity of the adjacent fatty tissue.

Average of spinal cord.—Table 13 contains the analytical data obtained in regard to the spinal cords of the different animals. Besides the spinal cords proper, these samples included the layer of fatty matter which surrounds the spinal cord in the spinal canal. In some instances the quantity of material was not sufficient to make a determination of the ash, and in three instances the whole of the material was lost. The data show great variations in the composition of the spinal cords of different animals, especially in the content of fat and water. The Berkshire had a spinal cord in which the water predominated, while in Duroc Jersey, No. 7, the fat was the predominant constituent. The nitrogenous substances are not so large as would be expected in nerve tissue, and those which are present consist chiefly of the proteids and gelatinoids, the flesh bases being only in relatively small quantity.

Average of tendons.—Table 14 contains the analytical data for the tendons of the animals, with the exception of two cases where the samples were lost. Considerably more than half of the tendons in the fresh state is water, while the fat, as is to be expected, is quite low. The nitrogenous substances, next to the water, constitute the chief material in the tendons, showing the largest percentage of nitrogenous matters of any part of the animal, with the exception of the hoofs. The true proteids and gelatinoids constitute by far the largest portion of the nitrogenous substances, the flesh bases being in relatively smaller proportion. The ash in the tendons is higher than in the meats. The summation of the analyses shows uniformly more than 100 per cent, which is probably due to the use of too large a factor in computing the proteids of the different classes from the percentage of nitrogen. Variations in the composition of the tendons are sufficiently well shown in the footings of maxima and minima. The variation in the content of water is not great, while in fat the range is a very considerable one, as indicated by the percentages. The agreement in the percentage of nitrogenous substances is quite close, the tendons showing very little variation from a mean composition. The ash is also quite constant, the range of variation not being very great, except in the case of the Poland China.

Average of hoofs.—Table 15 contains the analytical data relating to the hoofs of the animals. The fat content of the hoof is extremely small, while water constitutes almost half the entire weight of this substance. The nitrogenous substances were not separated into three portions, but were all estimated as proteids by multiplying the nitrogen content by the factor 6.25. Considerably more than half of the total weight of the hoofs in the fresh state consists of nitrogenous material. The ash is not very high, only in one instance exceeding 1 per cent. The summation of the analyses shows in every case more than 100, except in the instance of the Duroc Jersey, and this is doubtless due to using the factor 6.25 in computing the total amount of nitrogenous substances, inasmuch as the factor for the flesh bases, which were not determined in this case, is considerably lower than the one just mentioned.

LOSS OF WEIGHT IN TRANSPORTATION.

Table 16 shows a comparison of the weights of the entire animal and the various cuts, as determined in Chicago and in Washington, showing the percentage of gain or loss in transit. The weights in Chicago presumably were made with great care, but were not controlled by any employee of the Division of Chemistry. The weights in Washington were made directly by the Division of Chemistry, and can be certified as absolutely correct. In five instances the weights ascertained in Washington were less than those ascertained in Chicago, and in three instances greater. The largest variation between the two weights was shown in the case of the Chester White, where the loss was 8.07 per cent of the whole weight. The smallest variation was found in the case of the Duroc Jersey, No. 6, with a loss of 0.27 per cent. The largest gain in weight was in Duroc Jersey, No. 5, namely, 1.18 per cent, and the smallest gain in weight was found in the Yorkshire, namely, 0.49 per cent. The table contains not only the total weight of the animal in pounds and grams, but also the weight of each cut.

RATIOS OF MEAT, BONES, ETC., TO TOTAL WEIGHT.

Table 17 contains the relative percentages of the different parts of the animals, excluding the head, leaf lard, and kidneys, which had been removed before shipping from Chicago. This table is of great practical and economical interest, showing the relative percentages of each constituent of the animal, based upon its entire weight. In the animals dressed as received by us it is seen that nearly 89 per cent of the total weight of the animal is meat (fat and lean), a little over 6.25 per cent bones, nearly 4.75 per cent skin, 0.16 per cent tendons, 0.12 per cent marrow, 0.08 per cent spinal cord, and 0.08 per cent hoofs. There is quite a remarkable agreement in the relative proportions of these different constituents in the different animals. For instance, the widest variation from the mean in the percentages of meat in the animals examined was, in round numbers, only 2 per cent, while in the case of the bones it was numerically no larger, although relatively the variation was very much greater. In the case of the skin also the variation was not very marked. In the minor constituents the percentage of variation is great, but the actual variation in the different animals small. In regard to bones, the largest percentage was found in the Tamworth, and the smallest in the Duroc Jersey, No. 6. These show the extreme variations, and indicate that the Tamworth has a much stronger skeleton, so far as shown by weight alone, than the Duroc Jersey, No. 6.

PERCENTAGES OF THE SEVERAL CONSTITUENTS.

Table 18 contains the percentages of the different constituents of the entire dressed animal, excluding the head, leaf lard, and kidneys. The data are most interesting from a practical point of view. It is seen that of the entire animals 36.43 per cent was composed of water, 49.67 per cent of fat, 10.46 per cent of nitrogenous matter, and 2.11 per cent of ash. It may excite remark that the percentage of ash in the animal is so small when it is remembered that the whole of the mineral matter of the bones is included with the ash, but by referring to the table of the analyses of the bones it is seen that only about 25 per cent of their total weight is mineral matter, the rest being composed of water and organic substances. The water and the organic substances are included in the other data, and the ash therefore expresses only the mineral matters of the animal, including not only the bones, but also the mineral matters of the other tissues. In regard to the nitrogenous substances, their proportionate division into three classes is of interest. It is seen that of the whole amount 8.12 per cent belong to the proteids insoluble in hot water, and 1.10 per cent to the proteids of a gelatinoid nature, while 1.14 per cent belongs to the nitrogenous bodies representing the flesh bases. From a nutritive point of view, the true proteids are the most valuable. The gelatinoids are highly nutritious, but on account of their smaller quantity do not have so high an economic importance from a nutritive point of view as the other proteids. The flesh bases have a lower nutritive value, but are prized in many cases on account of their ready absorption and their stimulating properties, being already in a state suitable for partial assimilation. The summation of the analyses as a whole is extremely satisfactory, only a little over 1 per cent of the total weight of the animal being unaccounted for in the actual data obtained.

Comparison of breeds.—In regard to the details of the various constituents, it is seen that the Berkshire leads all the others in the percentage of water, namely, 43.10. The smallest percentage of water is in the Duroc Jersey, No. 6, namely, 30.31. The largest percentage of fat is found in Duroc Jersey, No. 7, namely, 57.68, and the smallest in the Berkshire, namely, 40.46. Of the total nitrogenous substances, the largest quantity is found in the Berkshire, namely, 13.02, and the smallest in the Duroc Jersey, No. 7, namely, 8.96. It is evident from an inspection of the table that the meat of the Berkshire is better for the production of muscular strength, while that of the Duroc Jersey, No. 7 is best suited for the production of animal heat. The Berkshire meat would be best suited for the use of our army in Cuba, while the meat of the Duroc Jersey, No. 7 would be best suited for the miners of the Klondike. These remarks are made without any expression of opinion concerning the type as a whole, but only on the data obtained from the two animals. The examination of a large number of typical animals of each of the breeds would be necessary to establish a definite rule of that kind. It is fair to presume, however, that the single animal is to a certain extent typical, and therefore represents to that extent racial characteristics.

LECITHIN.

The determination of lecithin in meat products is accomplished, as has been already described, by an indirect method; namely, by the extraction of the lecithin with a mixture of ether and alcohol and the determination of the phosphorus in the extract. From the quantity of phosphorus determined the percentage of lecithin is calculated by factors based upon the percentage composition of the lecithin itself. The data given for the lecithin should be accepted with certain restrictions, based upon the difficulty of applying the analytical processes. In the extraction of the fat by ether a certain quantity of the lecithin is removed. If, now, the residual lecithin be determined in the undissolved matters, namely, the dry flesh, the quantity obtained does not represent fully the whole amount originally present, but rather the quantity present in the muscular tissue itself. Therefore, in case of the meats especially, the data must be accepted as showing the quantity of lecithin in the fleshy portions of the meat, and not the quantity originally present in the fleshy portions plus the fat. In the case of the marrow and spinal cord, another difficulty presents itself; namely, that there was not a sufficient quantity of the material on which to perform the whole of the analytical operations. Inasmuch as the ether extract comprises a large percentage of the whole weight of these bodies, it is evident that the determination of the lecithin in this extract represents approximately the quantity present in the original material. On account of the paucity of this material, therefore, the lecithin was determined in these cases in the ether extract alone. If, however, the quantity be desired for the whole material, it is evident that the data given are not sufficiently large.

PHYSIOLOGICAL IMPORTANCE.

From a physiological point of view lecithin is of prime importance. It is quite certain that this body forms the transition state between the phosphates of the animal body on the one hand and the mineral phosphates absorbed by plants on the other. In the growth of plants the mineral phosphates are converted, to a certain extent, into lecithin, which is found especially in the seeds, those of an oily nature predominating in lecithin bodies. In the consumption of vegetable foods by animals the lecithin doubtless plays an important function in being transformed again into a mineral compound, namely, the tricalcium phosphate of the bones. Other portions of the lecithin become assimilated in the tissues of the body, and especially in the brain, spinal cord, and marrow. In the consumption of animal products by other animals lecithin again plays an important rôle in nutrition, forming on the one hand the bony structure of the animal eating the flesh, and on the other being again stored as lecithin in the tissues above mentioned. The data given, therefore, in the foregoing analyses are of great importance not only from their scientific interest, but also in representing in a general way the distribution of the lecithin in the various tissues of the body.

DISCUSSION OF THE LECITHIN IN PARTICULAR SAMPLES.

Lecithin in the meat.—In Table 9 it is seen that the mean percentage of lecithin in the residue after extracting the fat from the meats is 0.23. Inasmuch as almost the whole of the lecithin of the meats is found in the muscular tissues, this represents pretty fully the whole amount present in the original sample. The quantity, however, of lecithin in the fat extracted by the ether must not be neglected if we are to consider the total amount present in the original samples. It is noticed that there is a considerable degree of variation in the percentage of lecithin in the different animals, the minimum quantity being found in the Duroc Jersey, No. 5, and the maximum in the Duroc Jersey, No. 7. It is evident, therefore, that this variation is not to be ascribed to the influence of breed alone.

Lecithin in the bones.—The quantity of lecithin in the bones is considerably greater than that found in the meats, the mean being 0.31 per cent. In one instance, namely, the Duroc Jersey, No. 7, the lecithin was determined both in the residual bony matter and in the fat which was extracted. A great difference is noticed in the distribution of the lecithin among the various animals, the maximum quantity being found in the Poland China and the minimum in the Tamworth.

Lecithin in the marrow.—The quantity of marrow was so small that the only possibility of determining the lecithin was in the original ether extract. The data, therefore, are not as reliable as those ascertained by determining the lecithin in the extract after removal of the fat. In each instance the amount of lecithin was very small, except in the case of the Berkshire, where it was quite high.

Lecithin in the skins.—The mean quantity of lecithin in the skin was found to be 0.19; the maximum being 0.41 and the minimum 0.06. In three instances the lecithin was determined in the samples both after extracting with ether and in the ether extract. These cases are appropriately marked in the analytical tables.

Lecithin in the spinal cord.—Lecithin in the spinal cord was determined only in the materials extracted by ether. As was to be expected, the quantity is very high; the mean percentage being 1.54, the maximum 2.95, and the minimum 0.70. On account of the small quantity of the material it was not possible to determine the lecithin in the residue after the removal of the fat. If this could have been determined it is evident that the quantity of lecithin would have been very materially increased.

Lecithin in the tendons.—In one instance, namely, the Berkshire, the determination was made both in the extracted fat and the residue. In this case the quantity of lecithins is quite high. The mean for all the tendons, as determined, was 0.19, with a maximum of 0.45 and a minimum of 0.08.

In Table 18 the total percentages of lecithin in the whole animal, with the exceptions noted in several of the tables, are found. The mean percentage is 0.23, the maximum 0.42, and the minimum 0.11.

In submitting the above discussion it is but just to state that at the commencement of the analytical examination it was not our purpose to determine the lecithin at all. Had it been so, the determinations would have been made in a somewhat more satisfactory manner. The data, however, as submitted are, nevertheless, valuable, and with the restrictions noted in the different tables may be relied upon as a basis for economic studies.

CONCLUDING OBSERVATIONS.

In conclusion it may be stated that although work of the kind which has just been discussed is extremely onerous and time-consuming, yet it appears from a study of the results obtained to be a further contribution to our knowledge of dietetic science. All systems of true dietetic studies must rest first of all upon well-established chemical data. No valuable conclusions in regard to the dietetic value of any food can be obtained without first having ascertained its exact chemical composition. This having been done, the further study of its dietetic value rests also upon its chemical properties, as, for instance, the coefficients of digestibility. It appears advisable, therefore, considering the character of the data which have been presented, to recommend that studies of this kind be continued with all the classes of animals used as foods in this country. It would be advisable, if possible, that in studies of this kind, the animals be slaughtered at or near the point where the chemical examination is to be made; or, if this be not convenient, that a representative of the Chemical Division be present at the time of the slaughtering for the purpose of ascertaining the quantities of blood, hair, and excreta from the different animals and obtaining representative samples thereof for chemical examination.

Our systems of feeding and our environment develop types of animals which are quite distinct from those grown in other lands, and therefore the data which are obtained on animals in other countries are not strictly applicable to studies of the economic science of food production and food composition in this country.

APPENDIX.

For full particulars relative to the general principles of the separation of the different forms of nitrogenous bodies the reader is referred to the Principles and Practice of Agricultural Analysis, volume 3, and to Bulletin No. 54 of this Division. An abstract of the literature relating to the separation of flesh bases from other nitrogenous bodies is given here.

PRECIPITATION OF PROTEIDS SOLUBLE IN WATER BY CHLORIN AND BROMIN.[1]

Rideal and Stewart recall some of the experiments made in 1876, in which it was shown that a current of chlorin gas conducted through an aqueous solution of proteid matters produces a precipitate which is of a quite constant composition, and one which can be collected, dried in vacuo, and weighed. They describe particularly the use of this reagent in precipitating gelatin prepared from the high grade commercial article. They show that the total quantity of gelatin can be accounted for from the weight of the precipitate by multiplying the weight of the precipitate obtained by the factor 0.78. The authors also point out the possibility of using bromin in place of chlorin for the precipitation, and state that the studies of the use of bromin are under way. They call attention also to the fact that as early as 1840 chlorin had been used by Mulder for the precipitation of soluble proteids, and refer to a paper of his published in Berzelius's Jahresbericht, volume 19, page 734, in which he obtained results on precipitation quite similar to those secured by Rideal and Stewart.

Other references to the literature on the subject are also given, viz: De Vrij, Ann. Pharm., lxi, 248; Thénard, Mém. d'Arcueil, ii, 38; Mulder, Bulletin en Néerlande, 1839, 153; and Berzelius' Jahresbericht, xix, 729.

Allen and Searle, acting on the suggestion of Rideal and Stewart, worked out the bromin method by applying it to various soluble proteids, including the whole range from albumin to peptone. In the application of this test to commercial gelatin 50 grams of commercial gelatin are dissolved in warm water and the solution diluted to half a liter. In 10 c. c. of this solution, corresponding to 1 gram of the gelatin, the nitrogen is determined directly by the Gunning-Kjeldahl process.

Another portion of 10 c. c. is treated with an excess of bromin. The solution is first brought to a volume of 100 c. c. with water and placed in a conical beaker with a sufficient quantity of hydrochloric acid to produce distinct acidity. A saturated solution of bromin water is added in considerable excess and the liquid stirred vigorously for some time. The precipitate which separates is flocculent when first formed, but becomes more viscous after stirring and adheres for the most part to the sides of the beaker, which, with its contents, is allowed to stand for about half an hour, or until all the precipitate is settled. The supernatant liquor is decanted through an asbestos filter. The precipitate adhering to the beaker is washed several times with cold distilled water and the washings poured through the filter. Occasionally, when most of the free bromin is washed out of the precipitate, the liquid

[1] The Analyst, 22, pp. 228 and following; also pp. 255 and following.

does not filter clear. It is therefore advisable to keep the washing separated from the filtrate, and, if necessary, wash with sodium sulphate solution or with bromin water. The nitrogen in the precipitate is determined by the Gunning-Kjeldahl process as follows:—

The precipitate which has been collected on the asbestos filter, together with the asbestos, is returned to the beaker in which the precipitation took place. Twenty cubic centimeters of strong sulphuric acid are added, the beaker covered with a watch glass and placed on a wire gauze over a lamp. When frothing has ceased about 10 grams of powdered potassium sulphate are added and the liquid boiled until colorless. After cooling it is diluted with water and the ammonia distilled off and determined in the usual way. The percentage of nitrogen found, when multiplied by the factor 6.33, or, in the case of gelatin, by 5.5, gives the amount of proteid matter precipitated by bromin. In the commercial gelatin above mentioned the nitrogen content was found to be 14.1 and 14 per cent, respectively, on two determinations. Solutions of creatinin, asparagin, and aspartic acid were found to yield no precipitates with bromin, but bromin was found to precipitate all albumin, acid albumin, and all peptones formed by the digestion of albumin with pepsin.

NITROGEN IN MEAT EXTRACTS.

On applying the bromin method to commercial meat extracts the following results were obtained. The solutions of the Bovril preparations were not previously filtered and therefore the figures contain the nitrogen in the fiber present:

Relative amounts of nitrogen in meat extracts.

	Nitrogen in precipitate by bromin.	$N \times 6.33 =$ proteids.
	Per cent.	*Per cent.*
Liebig Company's extract	1.41	8.92
Seasoned bovril	1.94	12.28
Bovril for invalids	2.64	16.71

Koenig and Boemer have shown that the proteid nitrogen in meat extracts is generally much overestimated. They found a total of 1.17 per cent of proteid nitrogen in the Liebig Company's extract, which is equivalent to 7.41 per cent of total proteids, mostly albumose.

PROBLEMS SOLVED BY THE BROMIN METHOD.

The fact that bromin completely precipitates all proteid and gelatinoid matters in solution affords a convenient means of solving certain problems which have hitherto presented considerable difficulty. For instance, in a solution which has been subjected to digestion it may be possible to precipitate all the unchanged proteids by saturation with zinc sulphate. The peptones which have been formed during digestion remain in solution and can be separated by filtration. In the filtrate the peptones can be completely precipitated by bromin, and thus the total quantity of these bodies formed during digestion can be accurately determined.

Allen and Searle applied this method to an examination of the Liebig Company's extract, 5 grams of which were dissolved in 100 c. c. of water and the solution saturated with zinc sulphate. After filtering, bromin water was added to the filtrate and a precipitate produced which redissolved on diluting with water and the addition of hydrochloric acid. When the filtrate from the saturated zinc sulphate was previously diluted with water and acidulated no precipitate was formed on the addition of bromin. This reaction shows that no considerable quantities of real peptones exist in Liebig's extract.

www.ingramcontent.com/pod-product-compliance
Lightning Source LLC
Chambersburg PA
CBHW020329090426
42735CB00009B/1464